Digital Transmission Systems and Networks

ELECTRICAL ENGINEERING, COMMUNICATIONS, AND SIGNAL PROCESSING

ISSN 0888-2134

Raymond L. Pickholtz, Series Editor

ANOTHER WORK OF INTEREST

Local Area and Multiple Access Networks
Raymond Pickholz, Editor

---

*These previously published books are in the Electrical Engineering, Communications, and Signal Processing series but they are not numbered within the volume itself. All future volumes in this series will be numbered.

Volume I: Principles

Digital Transmission Systems and Networks

MICHAEL J. MILLER
The South Australian Institute of Technology

SYED V. AHAMED
The City University of New York

COMPUTER SCIENCE PRESS

Computer Science Press
1803 Research Boulevard
Rockville, Maryland 20850

1 2 3 4 5 6 Printing   Year   92 91 90 89 88 87

Library of Congress Cataloging-in-Publication Data

Miller, Michael J., 1939–
  Digital transmission systems and networks.

  Bibliography: p.
  Includes index.
  Contents: v. 1. Principles — v. 2. Applications.
  1. Digital communications. I. Ahamed, Syed V.,
1938–   . II. Title.
TK5103.7.M55   1987   621.38   86-33357
ISBN 0-88175-094-8 (v. 1)
ISBN 0-88175-176-6 (v. 2)

Dedicated to
Edith, Kristina, David, Karen
and
Ameera, Sonya, Nisha

VOLUME I
CONTENTS

VOLUME II
CONTENTS

PREFACE

With the rapid growth of digital techniques in telecommunications networks, there is a need for a revised approach to the study of the principles and the applications of digital transmission systems and networks. The purpose of this two-volume text is to provide an introductory treatment of principles involved in digital communications (Volume I) and to integrate these principles into the applications environment (Volume II). The engineering aspects of integrated communications systems for telephony, television, data, and other network services are discussed in the second volume.

These texts should be useful for senior undergraduate students in the communication sciences, graduate students, and practicing engineers. They should also prove valuable to computer science students and software designers who wish to understand networks and their control. Network control and the design of special purpose software, associated with the management of networks under normal and abnormal conditions needs a firm grasp of the principles discussed in Volume I and their applications discussed in Volume II.

The two volumes are intended for a sequence of two semester courses in digital transmission systems and networks. Some familiarity with the principles of analogue communication systems will give the reader an appreciation of the digital techniques discussed in the two volumes. Prior exposure to basic probability theory related to random processes would be helpful.

Volume I, in particular, is concerned with the evolution of the digital networks, the different types of signals encountered in baseband digital transmission, intersymbol interference and pulse shaping, signal regeneration, measurement techniques, and digital encoding of speech. Considerable effort has been made to make the material useful to senior undergraduate students and to practicing engineers in the telecommunications industry. Emphasis is therefore given to a careful presentation of the essential concepts and typical engineering solutions with computer programs and without becoming unnecessarily concerned with too many theoretical proofs. The rapidly changing VLSI environment preempts our detailed presentation of the hardware realizations of the basic building blocks of these baseband systems.

Volume II is concerned with implementation of the principles (discussed in Volume I) in the physical networks for the transmission of data. In particular, digital radio, telephone, and computer networks are presented together with the implementation of error control in such networks. Integrated Services Digital

Networks (ISDN) and Digital Subscriber Systems (DSS) are also presented in considerable detail in the light of their capabilities, design, optimization, trade-offs, and the potential impact on the telephone networks around the world. We have delineated the steps and the optimizations undertaken in the early designs and progress of the 144 kbit/s facility and the 56 kbit/s circuit switched digital capability introduced by AT&T in the early eighties as a precursor to the ISDN-like network services now being offered by certain telephone operating companies in the United States. This volume is directed towards graduate students in tele-communications and practicing network and design engineers. It also provides an insight into the devices necessary to realize the network terminal functions. System programmers and software designers will find these discussions attractive in implementing the network control strategies.

The emphasis in these texts is on a pedagogic approach to understanding essential engineering concepts. Often this is achieved by examining a number of specific cases which illustrate an idea rather than by attempting to develop proofs for the general case. For many practicing engineers several years out of graduation, some basic tools of mathematical analysis have lain dormant and need to be revitalized. For a professional engineer working in a rapidly developing field such as telecommunications, an analytical appreciation of new ideas is essential. The purely descriptive approach is too limiting. The use of many problems with worked solutions in the text is intended to foster the reader's understanding and development of these analytical skills.

We anticipate that these texts should fill a gap in the current literature on the subject particularly with its emphasis on evolving digital networks. It can provide the basis of courses for senior students and for professional development courses for practicing engineers.

We are particularly grateful to Telecom Australia for supporting the initial preparation of this material. The result reflects the benefits of practical experience gained by each of us as employees of Telecom Australia and Bell Telephone Laboratories Inc., respectively. It is almost certain that this work would never have been commenced were it not for John Grivell, a senior engineer in Telecom Australia, who attended one of our lecture courses and recorded and annotated lecture notes which were much more clear and comprehensive than the original source material. We also wish to sincerely thank Dr. Teresa Buczkowska of the New South Wales Institute of Technology for contributing the chapter on computer networks and Professor Shu Lin of Texas A&M University for his influence on the section on error control. Thanks also go to Elaine Milsom, Judy Duval, and Isobel Keegan for their assistance with typing.

We also thank President Edmond L. Volpe, and Dr. Barry Bressler, Vice President for faculty and instruction at the College of Staten Island, City University of New York, for their support and encouragement. The authors are also indebted to the management of Morris Research and Engineering Center of the

Bell Communications Research at Morristown, New Jersey, where the original research work from AT&T Bell Laboratories was considerably extended. The opportunity to carry on the basic ISDN and CSDC work was initiated under the direction of Eric E. Sumner of the AT&T Bell Laboratories and by Frederick T. Andrews, now of Bell Communications Research. Dr. Ralph W. Wyndrum, Dr. Barry Bosik, Dr. Harold Seidel, and Dr. Peter Bohn of AT&T Bell Laboratories have also influenced the direction and findings we present in the last two chapters of the second volume. Dr. N. S. Jayant, William L. Shafer, and Albert J. Schepis of the AT&T Bell Telephone Laboratories, Joseph F. Urich, Rein R. Laane, Dr. Richard A. McDonald, Wilhelm H. von Aulock of Bell Communications Research have provided constructive comments based upon the manuscript we had presented to them.

Michael J. Miller,
South Australian Institute of Technology

Syed V. Ahamed
The City University of New York

July 1986

Chapter 1

THE DEVELOPMENT OF DIGITAL NETWORKS

1.1 INTRODUCTION TO DIGITAL TELECOMMUNICATIONS

1.1.1 The Digital Revolution

During the past decade, a revolution has been quietly but persistently taking place in the telecommunications industry. For most of this century, analogue methods have dominated the multiplexing and transmission techniques for telephony, television, and related services. Now, after many years of research and development, the practical implementation of digital communications techniques has come into being and is growing at a rapidly accelerating rate worldwide.

In Japan, for example, high-capacity digital transmission networks have been developing rapidly. The Nippon Telegraph and Telepone (NTT) public corporation currently has a transmission line digitization program designed to have 90 percent of all long distance circuits digital by the mid-1990s (1).

In the United States, a major growth area of digital transmission has been in interconnecting pulse code modulation (PCM) short-haul cable carrier systems, known as T1 carriers, between telephone switching centers. Introduced in the early 1960s, there were about 2 million channels in service by the 1970s (2). Projections indicate that in the U.S. network, considerably more digital than analogue transmission systems will be installed per year by 1990. As the century nears its end, it is expected that almost all new system additions will be digital.

In the United Kingdom, and in many European countries, significant developments have been taking place in the introduction of PCM terminals and digital transmission systems. Italy, for example, had over five hundred 13 GHz digital radio terminals in operation by 1980.

In countries with large distances between regions of high population density, the new emphasis on digital telecommunications is reflected in the development of new long-haul digital transmission networks. In Australia, a transmission system is being installed that incorporates thousands of miles of digital microwave radio systems. This system is capable of 140 Mbit/s transmission rates to inter-connect the main population centers. In Canada, an 8 GHz digital radio system is being developed from coast to coast.

Some of the reasons for the accelerating development of digital transmission systems are as follows:

(1) *Developments in micro-electronics*, especially in Very Large Scale Integra-tion (VLSI) technology, have enabled more economic manufacture of digital multiplexing and transmission equipment with high reliability and increased capability.

(2) *Developments in data networks* are necessary to support the rapid growth in the "information explosion" associated with computer networks for data processing, intra-corporate data flows, electronic mail, computerized bank-ing and booking agencies, and international library networks.

(3) *Developments in telecommunication networks* can provide additional cus-tomer facilities through the added flexibility and transmission performance benefits obtained in an integrated services digital network.

Although digital transmission and switching technology was relatively slow in birth, it is now evolving with explosive acceleration. The influence of digital techniques in communications is sweeping forward like a tide. It is being born along successively on waves of change in allied fields including microelectronic device development and the emergence of new transmission media. It is being fueled by the increasing demands for new types of telecommunication services. Its progress is being fostered by the availability of new computer hardware and software technology.

Digital transmission techniques are rapidly being introduced in most countries. Having embarked on this development, it is natural that digital switching systems should evolve in parallel. The most daunting and time-consuming task in this work has been the design of the computer control equipment and the writing of the necessary software programs for control of the solid-state switching functions. A large range of digital switching systems is now available, varying from major 100,000 line local exchange switching units down to small private exchanges for business purposes.

Now a new evolutionary phase is on the horizon. The growing desire to provide and integrate digital telecommunications services of all types has led to the development of the concept of the Integrated Services Digital Network (ISDN).

This reflects the fact that digitally encoded information, be it voice, data, video, text, graphics, or network control information can all be handled in a uniform way. This leads to the possibility of the integration of services within a single network structure.

Furthermore, individual users are demanding that it be possible to have parallel access to these services via their private lines. They wish to be able to conduct a telephone call from their homes and simultaneously send or obtain data from an electronic mail center or other data base. Mixed parallel communication is a natural requirement for the human being who is used to communicating simultaneously by hearing and sight.

Therefore, in addition to the conventional telephone service, the telecommunications user of the future can expect to have available a wide variety of mixed additional services with an ISDN. These services will be available in parallel from a new subscriber line termination which will give access to worldwide networks with high standards of performance.

The work of developing concepts and standards for the ISDN began in earnest in 1978 in the international telecommunications forums. Since then great advances have been made. The ISDN will provide for digital multi-purpose connections extended to the users' premises. This will permit mixed communications including speech, text, and data. The ISDN transmission and switching channels will be through-connected digitally from end to end. At the user's premises, speech, text, or data terminals may be connected in parallel to the subscriber line.

The worldwide development of the ISDN will be a long-term one. The time scale for its achievement, at first nationally and later internationally, will depend on the rate at which digital transmission and switching systems can be implemented.

A possible scenario for these developments is suggested in Figure 1.1, which is adopted from Iwasaki (1). It shows three stages. Stage 1 represents a period when the individual telephone networks and data networks are progressively digitalized. By the 1990s it is predicted that the majority of individual service networks will be primarily digital. There will probably be some analogue transmission links and switching offices still in operation but PCM systems will have become dominant.

In Stage 2, a number of new nontelephone services will have become integrated to meet the demand for diversified communications. This will include the integration of circuit switched and packet switched data services, together with teletex and facsimile services. At this stage, it can be expected that work will have begun on developing the user-to-network interface systems and the exchange systems for the ISDN.

Stage 3 is the period in which the telephone networks and data networks will be integrated. The fully-implemented ISDN will make its appearance for telecommunications users in the next decade.

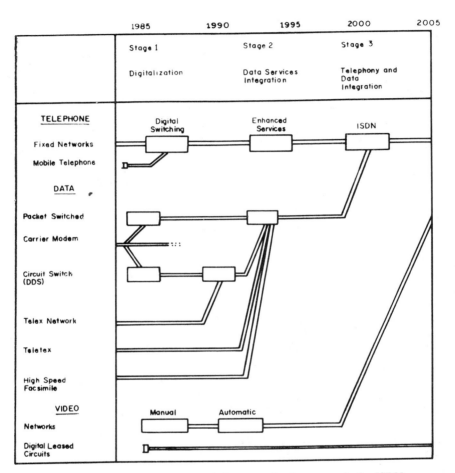

Figure 1.1   Development of digital communications towards the ISDN.

1.1.2   Digital Network Development

In a telecommunications system, the greatest utilization is by telephone subscribers. In terms of bandwidth considerations, television relay requirements also
represent a major demand. Both of these applications require the transmission
and switching of analogue or continuous waveforms. As a result, *analogue
communication technology* has dominated up to the present.

Analogue telecommunications technology is characterized by frequency-division-multiplex (FDM) techniques for carrier transmission and electro-mechanical space-division-multiplex (SDM) for switching. Although computer systems

have been more recently introduced for control of telephone exchanges (switching centers), the switching techniques for interconnecting circuits have still been electromechanical. These include the use of miniature reed-relays or some similar electromechanical switch.

In an analogue network, binary signals associated with data transmission must usually be converted into analogue form using voice-band modems (modulator/demodulators). By use of frequency-shift-keying (FSK), phase-shift-keying (PSK), or other modulation methods, the data signal is converted into a form suitable for transmission within one speech channel of an FDM carrier system.

In a *digital communications system*, the relative influence of continuous signals, such as speech, and data signals are reversed. In a way, it is almost as though "the tail is wagging the proverbial dog." The transmission channels and switching techniques are designed primarily to ensure reliable connection of signals in binary form.

Digital communications systems are characterized by the following:

(1) pulse code modulation (PCM) for sampling, quantizing, and coding telephone signals using eight bits per sample
(2) time-division multiplexing (TDM) for combining coded speech, video, data, and other signals for transmission
(3) digital switching in which switching between different time slots on different TDM buses is achieved by a combination of space switching by electronic digital gates, and time switching using electronic memories
(4) digital transmission in which the composite baseband digital signal is transmitted over cable, radio, fibre, or other transmission media.

In terms of demand on transmission resources in the network, the important categories of services are speech, video, and data. Here "data" is taken to include computer communication, electronic mail, facsimile, and telex.

For the next decade at least, the transmission requirements of the speech service are expected to far exceed the combined requirements of all other service categories. This is particularly true in switched services because the great majority of telephone traffic is switched, whereas a significant proportion of the video and data service demand will probably involve high bit-rate traffic handled on a nonswitched basis.

It has been estimated (3) that until the year 2000 the transmission requirements in the network for all data services is not expected to exceed one percent of the digital telephony transmission requirement if present pulse code modulation techniques are retained. However, the transmission requirements of the speech services will probably be compressed by a factor of up to 8 or 10 by the use of processing and interpolation. In that event, it is predicted that the total transmission requirement for data could account for as much as 10 percent of the telephone transmission requirements by the year 2000.

Within the data communication category, it is predicted that in the timeframe 1990 to 2000, computer communications will account for 80 to 90 percent of the total traffic volume. The balance of data traffic will probably be divided equally between facsimile and electronic mail, at least for those countries that use a phonetic alphabet.

For video transmission, the resource requirements are likely to vary widely from one segment of the network to another. By the year 2000, it has been predicted that video transmission requirements for television broadcasting and teleconferencing may total 20 to 30 percent of the telephone transmission requirements on some routes.

It seems clear that the speech service will continue to dominate the transmission system volume requirement. Therefore, it is necessary that any planning for a new digital network infrastructure be justified in terms of its relative advantages in respect to the telephone service.

1.1.3 Pulse-Code-Modulation Fundamentals

Pulse code modulation (PCM) concepts were established in the late 1950s and have become the most common technique for transmission of speech in a public network. As illustrated in Figure 1.2, PCM is characterized by five processes. These are sampling, multiplexing, quantizing, companding, and coding.

There are two main classes of PCM systems. A 24-channel PCM system is used in the U.S. and Japan. In Europe, Australia, and many Asian countries, a 30-channel PCM system is used. In the 30-channel system, each of 30 telephone

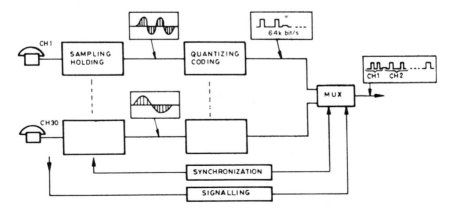

Figure 1.2 Pulse code modulation and time division multiplexing.

channels is sampled 8,000 times per second. Each analogue sample value is then encoded using eight binary bits per sample.

There are a total of $2^8 = 256$ possible discrete amplitudes, which can be represented by any one 8-bit code word. Therefore, before encoding, it is necessary to *quantize* the PAM sample amplitudes by subdividing the amplitude range into 256 small intervals. The truncation error in the quantizing process is known as *quantizing noise*.

The quantizing noise power can be reduced by using unequal quantizing intervals with small intervals for small-amplitude voltage values and larger intervals for large-amplitude less probable sample voltage values. This is known as *compression*. At the receiving end a matching expandor is required. The combined processes are referred to as *companding*.

After sampling and encoding, each speech channel signal is represented by a 64 kbit/s binary stream. This is discussed in greater detail in Chapter 6.

The 64 kbit/s bit streams from each of the 30 channels are then time-division multiplexed together with two additional 64 kbit/s bit streams. One of these two streams is used for synchronization. The other contains the signalling associated with the 30-channel multiplexed group. The resulting composite time-division-multiplex pulse code modulation (TDM-PCM) serial binary 30-channel signal consists of frames 125 μsec long. This is illustrated in Figure 1.3. The length of a frame is equal to the sampling interval. Each frame contains 32 8-bit binary code words. Of these, 30 are used for the 30 speech-channel samples. They are the time slots 1-15 and 17-31. The first time slot in each frame contains a special 8-bit sequence used for frame synchronization at the receiving end of the PCM system. More details on frame synchronization are given in Volume 2, Chapter 2. The 8-bit sequence in time slot 16 conveys dialing and signalling information.

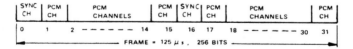

Figure 1.3   30-channel PCM frame structure.

Exercise 1.1

For the 30-channel PCM system described above, determine the transmission bit rate.

Solution. Each frame is 125 μsec long and contains 32 8-bit code words. Therefore, the bit rate is

$$R = 32 \times 8 \times \frac{1}{125 \times 10^{-6}} = 2048 \text{ kbit/s.}$$

Note that this is equivalent to 64 kbit/s per speech channel.

Figure 1.4 illustrates the frame structure for the 24-channel PCM system used in the United States and Japan. Each voice channel is sampled 8,000 times per second as in the 30-channel primary multiplex system. Also each sample value is represented by a code word of 8 binary bits as in the 30-channel system. However, every sixth frame, the least significant bit in each code word is "stolen" for signalling. That is, the last significant bit is deleted and replaced by a signalling bit.

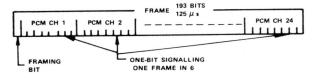

Figure 1.4   24-channel PCM frame structure.

In the Bell T1 PCM system, each frame of 24 8-bit words is preceded by a single bit which is used for frame synchronization. As a result, the transmission bit rate for the 24-channel system is

First-order multiplex (24 ch.) bit rate $= 8000 \, (24 \times 8 + 1)$ bit/s

$= 1544$ kbit/s.

The British also use a 24-channel system. It does not use an additional digit for frame synchronization. A further level of time division is involved in the method of frame synchronization and signalling that is used. A group of four frames is called a multiframe, whose repetition frequency is thus 2 kHz. The first digit of each channel in frames 1 and 3 is used to convey line signalling information. The first digit in frame 2 is available to provide a data transmission channel capable of transmitting at a rate of $2 \times 24 = 48$ kbit/s. The first digit in channels 9-24 of frame 4 convey the synchronizing signal.

In some PCM systems such as the U.S. Bell T1 carrier system, the various functions, including sampling, encoding, and multiplexing, are divided between "channel unit" circuit subassemblies (one printed-circuit card for each voice channel) and "common equipment" cards. The circuits make use of VLSI technology. Each channel unit is the interface between the incoming analogue signal and the common equipment subassemblies. The channel unit performs the following functions:

(1) two-wire to four-wire conversion using a hybrid transformer
(2) detection and generation of appropriate signalling information (for example, on-hook off-hook transitions, or dial pulse)
(3) provision of exchange battery to subscriber wire pairs
(4) filtering, sampling, and encoding incoming analogue signals using per-channel codecs (coder/decoder units)
(5) decoding and filtering the received binary voice signals
(6) interfacing the 64 kbit/s PCM signal to the common equipment subassemblies

The common equipment subassemblies accept the PCM signals and signalling information from the channel units and perform the following functions:

(1) generation of 8 kHz and line transmission rate clock signals
(2) digital multiplexing of 64 kbit/s streams into a TDM sequence
(3) addition of frame synchronization bits
(4) conversion and multiplexing of signalling information in binary forms and incorporation into the line transmission signal

At the present time, PCM techniques are chosen almost universally for digital transmission of speech in public networks. Each speech channel generates 64 kbit/s. In Chapter 6, methods for analyzing the performance of PCM systems will be examined. Also, system design considerations in relation to international standards will be discussed.

There are other important methods used in public telecommunication networks for digitizing speech signals including Adaptive Differential Pulse Code Modulation (ADPCM) and Delta Modulation. Typically, these require 32 kbits/s per voice channel. Standards for the use of 32 kbit/s ADPCM speech transmission are also currently being developed. For more details, see Chapter 6 of this volume.

Other important speech encoding techniques are also being developed, which lead to even further reduction in transmission bit rates. In military systems, for example, where transmission quality may not be required to be as high as in public networks, techniques have been developed for transmitting speech at rates as low as 2 kbit/s per voice channel. The relative performance of techniques for encoding of analogue signals will be discussed in Chapter 6.

1.1.4 Growth in Data Transmission Demands

In many countries, the growth of digital transmission and switching technology has been accelerated by decisions to establish public digital data networks. In some countries, alternative types of public data networks have been put into service. Two such types of service which have become popular are as follows:

(1) a packet switched data service which provides for the transmission of data information in segments of a convenient size, called packets. Packets are sent between users by adaptive routing rather than by reserving a single physical circuit

(2) a dedicated leased line digital data service (DDS) which offers customers point-to-point and multiplexing options. For example, data from a single customer's several terminals in different locations can be transmitted by the network and multiplexed into a single 48 kbit/s stream for input to a central computer.

These types of data network services are discussed further in Volume 1, Section 1.3 and in Volume 2, Chapter 3.

The historical development of telecommunications in the past has been such that there are separate, optimized special purpose networks for the various communications services. Hence, separate networks have evolved for speech, telex, data, and video. In many cases, they share common transmission carriers.

Only the telephone and telex networks have developed into global open networks in which every user can access any other user worldwide. In the case of the younger data communications networks, few global networks have as yet evolved. This has been due to such factors as the use of different data transmission speeds and different protocols. Nevertheless, most countries anticipate a rapid growth in demand for national and international data traffic.

In the future, the Integrated Services Digital Network (ISDN) will replace the conventional telephone network. It will be a worldwide open network and seems destined to boost the development of new and efficient nonvoice services in combination with the voice service. Dedicated networks such as packet switched data networks will probably exist for some time beside the ISDN but eventually be incorporated within it.

1.2   THE COMPONENTS OF A COMMUNICATIONS SYSTEM

1.2.1   System Functions

It is not possible in a single text book or a single course of any reasonable length to attempt to discuss all aspects of digital communications systems. Any public communication system represents a vast and complex network of interconnected signaling, transmission, switching, and control equipment. In order to delineate those aspects that are treated in this text, it is necessary to first categorize some of the functions of a large-scale telecommunications network:

(1) *transmission* of information between pairs of points in a network. This function is taken to include multiplexing and, in the case of data networks, error control.

(2) *switching*, that is, the design of the topological structure of the network, and the methods for synchronization, switching, and routing.

(3) *signalling and control* which incorporates the facilities by which various signal sources such as telephone subscribers and data terminals establish, maintain, and clear calls through the network.

These functions are illustrated in Figure 1.5.

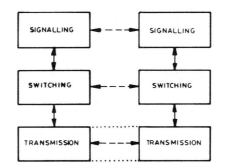

Figure 1.5   Structural layers in a communication network.

In the figure, complementary functions at either end of the transmission system are shown as separate layers in the diagram. The reader familiar with current developments in the design of protocols for data networks will also be aware of the use of the concept of separate layers or levels to separate classes of different functions (see Chapter 3 of Volume 2). It is interesting to note that this is not by any means a new technique unique to the computer network designer. Engineers have for decades tackled transmission system analysis and design using this layered approach. These functional levels will be examined as they apply to analogue and digital networks respectively.

1.2.2   Comparison of Analogue and Digital Telephone Systems

A simple example of an analogue type of communication system is illustrated in Figure 1.6. Before we examine this system, some terms need to be explained.

A *two-wire* connection is one in which the communication waveforms are transmitted in both directions along a single pair of wires. The term *baseband* is used to describe signals that are not modulated on a carrier but are simply in their original form as generated in the subscriber's premises.

In Figure 1.6, baseband signals from calling subscribers telephones are connected to a local exchange where they are switched to their destinations. A form of electro-mechanical switching is normally used to switch the signals from one

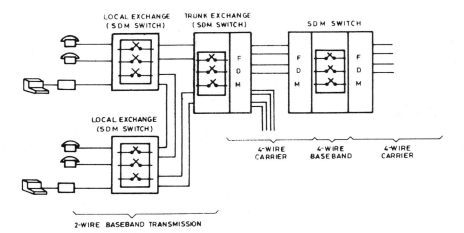

Figure 1.6   A network with FDM carrier transmission and SDM switching.

line to another. This switching is known as space-division multiplexing (SDM). The switches may consist of miniaturized relays operated under the control of a computer. This form of control is referred to as a stored program control (SPC) system.

If the circuit connection involves long-distance transmission, the subscribers and their associated 2-wire baseband local networks are connected to a trunk network. Switching centers known as trunk exchanges are used for this purpose. Analogue long-haul (toll circuit) transmission systems use frequency-division multiplexing (FDM). This permits the transmission of several speech signals in separate frequency slots over a single channel. In FDM carrier systems, the available channel bandwidth is divided into a number of nonoverlapping bands. Each speech signal is assigned one of these bands. Conventional analogue modulation and filtering techniques are used to place each speech signal into its allocated band and to extract it after transmission. A trunk exchange contains both SDM switches and FDM terminals as indicated in Figure 1.6.

When FDM carrier signals are to be switched at a subsequent exchange, they first have to be demultiplexed. This is because it is usually necessary to obtain the individual signals in baseband form before they can be switched by an SDM switch. Then if they are to be transmitted onto another trunk switching center, they have to be modulated and multiplexed again by FDM equipment.

More details can be obtained on analogue telecommunication networks in Flood (7); on switching systems in Pierce (8); on transmission in Bell Telephone Laboratories (9); and on signalling in Welch (10).

By way of comparison, a simple example of a digital communication system is shown in Figure 1.7. In the digital system, time-division multiplex pulse code

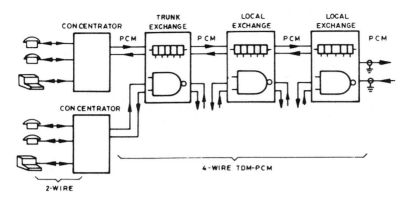

Figure 1.7 A digital network.

modulation (TDM-PCM) is used for multiplexing and transmission within the whole exchange network.

Subscriber circuits usually first enter remote switching stages known as concentrators which provide PCM conversion and multiplexing. More details are provided in Chapter 2 of Volume 2. Each concentrator is connected to and controlled by a local exchange. For switching in the local exchange, a combination of time switches T and space switches S is required. Signals must be switched from line to line (space switching) and from time slot to time slot in the PCM frame (time switching). Space switches use logic gates. Time switches employ shift register memories.

Digital transmission of the PCM bit streams between exchanges usually involves systems using telephone cable pairs for lower speeds and the shorter distances. For higher speed transmission, optical fiber systems are becoming popular. Alternatively, coaxial cable can be used. For high bit-rate long-haul transmission, microwave digitial radio systems are being rapidly installed in most countries. Digital radio techniques will be described in Chapter 1 of Volume 2.

In a digital network, the switching centers (trunk exchanges) associated with long-haul transmission perform essentially the same S and T switching functions as the local exchanges. The trunk exchange may be connected to subsequent trunk exchanges via a coaxial cable or fibre system.

One important aspect of the analogue and digital networks illustrated above is the number of modulation-demodulation steps involved. In the analogue system, a modulation-demodulation process is required at each switching exchange. This results in the accumulation of distortion and noise. By comparison, in the digital network, all the transmission and switching paths may be 4-wire PCM signals so that coding and decoding are only required when signals enter and leave the network.

The benefits to be gained by the implementation of digital techniques in telecommunications networks are as follows:

(1) *Digital circuitry* reduces hardware cost and improves reliability and maintainability. VLSI technology has reduced the failure rate of digital hardware by an order of magnitude. Digital systems with more than 100,000 components have been found to have measured down times of a few hours over periods as long as 20 years. Digital circuitry also permits the extensive use of automatic diagnostic techniques which reduce repair time and costs. Digital circuits save space through miniaturization.

(2) *Digital transmission* enhances the capability for high-volume signal distribution. The relative noise immunity of digital transmission can lead to better utilization of lines, fibers, and radio systems. Transmission error control techniques can also be applied. Digital transmission provides more readily for an integration of services as required by an ISDN. Digital transmission also permits better utilization of the subscriber line plant which has a very low usage rate in analogue systems.

(3) *Digital switching and processing* of communications signals can lead to more efficient services with a greatly increased level of sophistication. For example, a digital network can have a buffering capability. Digital information can be temporarily stored in low-cost memories without distortion. This may provide a variety of network benefits such as balancing of traffic load, speed conversion, and facilitating retransmission schemes for error control.

1.3 DIGITAL DATA TRANSMISSION

1.3.1 Voice-band Data Transmission

It was pointed out in the previous section that a primary factor in accelerating the development of integrated digital communications systems has been the rapid growth in quantity and diversity of data transmission services. Data transmission forms a much smaller component in any public communications system than the telephony component. However, society is becoming more and more dependent on highly sophisticated computer-based information exchange systems. Digital data transmission provides the means whereby these systems can be developed to meet rapidly changing demands.

Until recently, the majority of digital data traffic in public networks relies on the use of *voice-band data modems*. These are modulator/demodulators designed for transmission over conventional 300-3400 Hz analogue telephone channels. Data transmission requirements have represented a relatively limited traffic volume in most countries. Rather than constructing new separate digital transmission networks, it was more economical to make use of the existing analogue telephone network.

Private lines or lines through the public switched telephone network are used to provide the transmission medium. A private line is rented permanently or on a part-time basis by a subscriber. Before being placed in service, it is usually disconnected from the local exchange battery supplies and checked for its overall attenuation frequency characteristic to ensure a reasonable frequency response.

Modem connection via the public switched network is illustrated in Figure 1.8. A switched line through the network may be made up of a number of links chosen at random when the call is established. Although the frequency characteristics of individual links may be checked, it is difficult to predict the overall frequency response for the complete circuit. One reason for this difficulty is that significant mismatches may occur between individual links in the circuit.

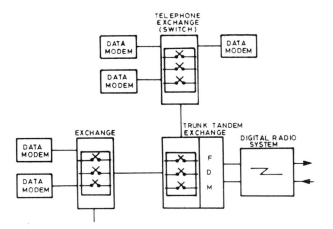

Figure 1.8 Data transmission over the switched telephone network.

The most common modulation techniques used in voice-band modems are as follows:

(1) *frequency-shift-keying* (FSK) for low speed modems for data rates up to 1200 bit/s. The line signal has a constant amplitude employing two tones (for example, 1300 Hz for a 0 and 1700 Hz for a 1) as illustrated in Figure 1.9.

(2) *phase-shift-keying* (PSK) for 2400 bit/s and 4800 bit/s, typically 4-phase PSK for the former and 8-phase PSK for the latter speed. For 4-phase PSK (sometimes written 4-PSK) the input binary data stream is subdivided into

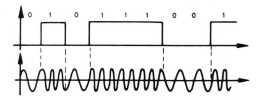

Figure 1.9   FSK line waveform (CCITT Recommendations V21 and V23).

two-bit blocks (dibits), each possible dibit pattern being represented by one of four phases of a transmitted tone as illustrated in Figure 1.10(a). The associated phasor diagram is shown in Figure 1.10(b) in which the length of each vector represents its amplitude and the angle shown indicates the relative phase. Usually the Gray Code is used in which adjacent phases differ in only one bit position to minimize the error rate caused by demodulation errors resulting from channel noise. For 4-PSK, a baud rate (symbol rate) of 1200 bauds provides an information transmission rate of 2400 bit/s.

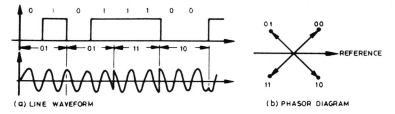

Figure 1.10   PSK signal (CCITT Recommendation V26 bis).

(3) *Quadrature amplitude modulation* (QAM) for data rates of 9600 and 14400 bit/s. For 9600 bit/s transmisison, 16-QAM is often employed. One of 16 phasors is transmitted in each symbol interval, each symbol representing four input data bits. Each phasor symbol represents a combination of phase and amplitude modulation. A typical 16-QAM phasor diagram is shown in Figure 1.11. Note from the phasor diagram that this is not a constant envelope signal as for the FSK and PSK techniques. For 14400 bit/s transmission, a 128-QAM signalling constellation is used. That is, each transmitted symbol is chosen from a set of 128 possible symbols.

If higher data rates are required, a group band FDM channel may be used which is equivalent to 12 voice channels. For example, a 48 kbit/s data stream modulated using single-sideband (SSB) modulation on a 100 kHz carrier may

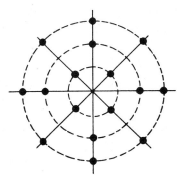

Figure 1.11 QAM signal phasor diagram (CCITT Recommendation V29).

be used to occupy the 60-108 kHz group band of an FDM carrier telephone system (CCITT Recommendation V36).

Problems in digital signal transmission over an analogue network: Although the voice-band analogue technique for providing data services conforms with the existing telephone-oriented public networks, as data traffic grows, the telephone network shortcomings become more evident and more costly to overcome. These include:

(1) *Reliability*—It is difficult to meet the reliability performance expected by data customers. As data rates increase, lower bit-error rates (say 1 in 10^7 or better) are expected but this may be difficult to guarantee.

(2) *Cost*—For higher data rates, modem designs must incorporate adapative equalization and become more complex.

(3) *Flexibility*—As the demand for more sophisticated information flow networks increases, it is difficult for voice-band services to provide for the flexibility and speed requirements. In many countries, concepts such as the "office of the future" and the "wired city" (16) envisage complex networks for data flow at rates from a few bit/s up to several megabit/s.

To meet these new demands, special purpose public data networks are now being developed in most countries.

1.3.2 Public Data Networks

A variety of data networks are being developed, some being national public networks provided by single national administrations and others being developed by independent telephone companies.

In Section 1.2.1, we discussed the three functions of any communications network, namely, transmission, switching, and signalling/control. The design of a data network requires the implementation of each of these functions in such a way that reliability, cost-effectiveness, and flexibility can be provided for a wide range of data transmission requirements.

In general, a public data network should have the following characteristics:

(1) *Baseband modems* are used wherever possible instead of the more complex carrier-type voice-band modems.

(2) *Time-division multiplexing* of low-rate data streams is used to generate higher rate composite signals. A transmission rate of 64 kbit/s for composite signals is commonly used. This is the same rate as that at which a single speech signal is transmitted over a PCM digital telephone system. For higher bit rates, 2048 kbit/s or 1544 kbit/s multiplexing ensures more efficient utilization of the transmission facilities of the network.

(3) *Robust digital techniques* specifically suitable for digital transmission are used in the user's loop termination and in the local and long-distance data links.

(4) *Strategical network planning* is used to develop a topology of nodes and interconnecting links which can provide high reliability and throughput with minimum delays and minimum cost. In the network design phase, analytical and simulation techniques based on queuing theory are used to optimize delay and throughput.

(5) *Packet switching or message switching techniques* may be used in data networks as an alternative to circuit switching used in speech communication networks. In a circuit-switched network, a complete circuit is used by two communicating parties during the whole period of information transfer. In contrast with speech communication using circuit-switched networks that have no appreciable transmission delay, data communication usually permits some delay in data delivery. Taking advantage of this fact, data networks often use store-and-forward switching. A message is divided into a number of shorter blocks called "packets" of the order of 1024 bits long. The packets each contain address and control information for transmission through the network independently of other packets, with routing dependent on the dynamic load on particular nodes and links. In message-switching, whole individual messages are transmitted independently through the network. Packet switching is a refinement of message switching in that smaller length packets are dynamically routed through the network so that delays are minimized.

(6) *Performance monitoring* of the data network is provided with automatically controlled standby facilities to guarantee good network error performance and availability. This is normally carried out from centralized offices. In-service performance monitoring, fault detection, and isolation are carried out at these operational centers using computer-based automatic testing and remote control facilities.

(7) *Error detection and correction techniques* are often provided, usually in the form of automatic-repeat-request (ARQ) retransmission systems. As a result, virtually error-free transmission (very high reliability) can be guaranteed.

(8) *A wide range of user data rates* are provided for, from as low as 50 bit/s to as high as several megabits per second.

(9) *Transparency of transmission* is provided so that the user has a "pipeline" available which can accept a wide range of data speeds and formats from synchronous and asynchronous sources. Although some formatting procedure for control purposes may be carried out at network access units located at the user's premises and within the network, this is not apparent to the user.

(10) *Full duplex facilities* are usually provided so that digital signals transmitted in each direction between nodes in the network usually represent a combination of user's data and network control, and addressing information.

In summary, the provision of a dedicated data network offers the user increased flexibility and better transmission reliability. In most cases, the operating costs for the user are reduced in comparison with those for comparable transmission rates in an analogue telephone-oriented network.

1.3.3 Nonswitched Data Networks

In several countries, two types of public data network services have been developed. One type is a nonswitched data network based on digital leased lines. The second type is a packet switched service.

Digital leased-line public data networks services were commenced in many countries in the past decade. These include the DATAROUTE network in Canada, the digital data service (DDS) in the U.S., Japan, and Australia. Similar networks have been developed in many countries for defense applications.

Figure 1.12 illustrates a DDS nonswitched public data network. It shows several customer offices interconnected via different paths through the digital data network. The network is shown consisting of a number of interconnected main centers to which the customers are connected via terminal centers. The customer's computing hardware is referred to as *data terminal equipment* (DTE). This may refer to a teletypewriter, card reader, line printer, laboratory sensor scanners, or a micro-, mini-, or main-frame computer. These are all incorporated by the term DTE, providing that they are equipped with a communication port.

The customer's interface to the DDS requires an interface unit known as a Network Terminating Unit (NTU) which converts the digital signals from a customer's DTE into a suitable format for transmission through the network. For example, the NTU may convert the 7-bit ASCII character set used by a teletypewriter into a format suitable for the network. A leased-line data network

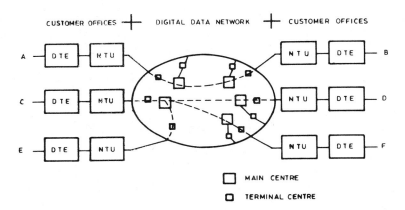

CUSTOMER OFFICES ── DIGITAL DATA NETWORK ── CUSTOMER OFFICES

☐ MAIN CENTRE

☐ TERMINAL CENTRE

Figure 1.12   The nonswitched Digital Data Service.

usually provides customers with synchronous full duplex data transmission at data rates of 2400, 4800, 9600 bit/s, or 48 kbit/s.

Users can usually arrange for their own network configuration in a DDS to provide one of several types of modes of connection. These include:

(1) *point-to-point*—as illustrated for terminals A and B in Figure 1.12.

(2) *point-to-multipoint*—as for terminals C to D, E and F.

(3) *multipoint-to-central-computer*—a configuration sometimes referred to as Netplex in which data from customers' data terminals at various locations will be transmitted by the network and multiplexed into a single high-speed 48 kbit/s stream for input to a central computer. A special Multiplexer Network Terminating Unit (MUX-NTU) is required at the central location. This is used to forward the necessary data, control, and timing information through a CCITT X.22 interface. See Tanenbaum (17) for details.

(4) a facility sometimes known as *Netstream* which allows separate customer data streams between DDS centers to be notionally aggregated for charging purposes to reduce costs. For example, a customer with five 9600 bit/s services between two cities will be charged for a 48 kbit/s stream. This will result in a lower tariff.

In a nonswitched data service such as DDS, if a link in the network fails, a standby link is automatically switched in to replace it. No other point-to-point error control procedure is provided.

1.3.4   Packet Switched Data Networks

The world's first packet switched data network was the Advanced Research Projects Agency Network (ARPANET), and has been in operation by the U.S.

Department of Defense since 1970. The first public packet switched data network was TELENET, which has been operational in the U.S. since 1975. This network was designed by Telenet Communications Corporation.

Other networks include the DATAPAC system of Canada, TRANSPAC of the French PTT Administration, the DDX system of the Nippon Telegraph and Telephone Public Corporation and AUSTPAC in Australia. These networks employ a standard access procedure known as *CCITT Recommendation X.25*.

As a result of this international standardization, there is the possibility of international networking between these systems. The EURONET, operating in nine European Economic Community (EEC) countries, is an example of international networking. ARPANET also now extends to nodes in Japan and other countries. It is planned that for many national public networks, access to some other national packet switched networks will be provided.

A packet switched data network is illustrated in Figure 1.13. Here, we consider only the main features. Packet switched networks are discussed in more detail in Volume 2, Chapter 3.

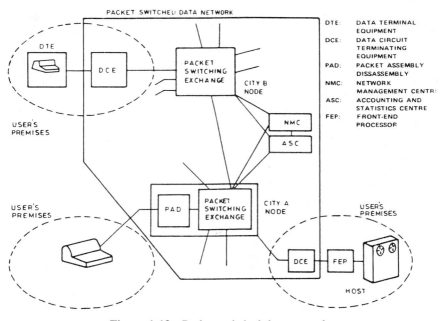

Figure 1.13 Packet-switched data network.

A customer's host or terminal with its interface is referred to as the data terminal equipment (DTE) as in nonswitched networks. The port of entry to the digital data network is called the Data Circuit-Terminating Equipment (DCE). This equipment is installed at the user's premises to provide all of the functions required to establish, maintain, and terminate a connection and for the signal conversion and coding between the DTE interface and the line. It may or may not be a specific or separate piece of equipment. Although the DCE is installed on the user's premises, it is considered part of the digital data network.

The DCE divides the users data into packets each consisting of about 1024 bits or less. In addition to the data message to be sent, each packet contains address information indicating its destination. In certain circumstances, this packet forming function is performed by a packet assembler/dissassembler (PAD). This is a special kind of concentrator which collects individual characters from asynchronous character-mode computer terminals and forms them into a packet.

The packet switched data network consists of a collection of computer-controlled switching nodes connected by digital communication circuits. The nodes are sometimes referred to as packet-switching exchanges.

A packet switching system is distinguished by the fact that the system adaptively routes individual packets rather than dedicating a channel for the whole duration of the message connection. As the packet enters the network at a switching node, that node examines the intended address and chooses a routing to the next appropriate node. At any particular point in time, the system may have several packets from the same source being routed to a single destination via different routes. At the receiver node, provision must be made for the storage, order, and checking of received packets so that they can be delivered to their destination in the correct order and without any losses or transmission errors. Protocol systems have been developed to make this possible. Protocols are sets of rules or conventions to which the operation of terminals and switching nodes must conform. They will be discussed in detail in Volume 2, Chapter 3.

Computer networks are designed and specified in a highly structured way. In an attempt to reduce the complexity and to ensure a logical and comprehensive approach, such networks are described in terms of a series of layers. Each layer is built upon its lower neighboring layer and provides certain services to higher layers. Each layer is associated with a well-defined subset of the functions to be performed by the network. The set of layers together with their protocols is called the *network architecture*.

The best known network architectural model is illustrated in Figure 1.14. It was proposed by the International Standards Organization (ISO) as a basis for the international standardization of network protocols. The two columns of 7-story layers in Figure 1.14 refer to the functions associated with the communicating DTEs. For example, these may be two mainframe computers. The two 3-story columns represent the communication system between the two DTEs. The double-headed arrows represent the protocols associated with that particular layer. A given layer on one computer "communicates" with the same layer on another computer.

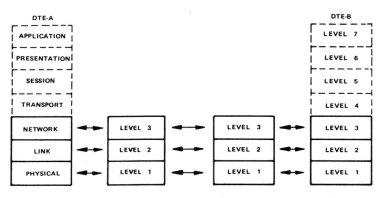

Figure 1.14   Architectural model for a computer network.

The following is a brief description of the functions of each of the seven layers in the ISO model, commencing at the bottom layer. More details are given in Volume 2, Chapter 3.

The Physical Level specifies the mechanical, electrical, and functional characteristics of the physical circuit by which a user's terminal is connected to the network. This defines how data bits can be exchanged across the DTE/DCE interface at a fixed rate with appropriate timing information. The physical circuit used to connect a user's terminal to the packet-switched network can be either an analogue or digital circuit as determined by the network provider. The data rates typically available on a packet-switched network and the respective interface requirements for packet mode terminals are summarized in Table 1.1.

Table 1.1
Network access method for packet mode terminals.

Data Rate	Type of Dedicated Network Access Circuit	Interface Requirement (CCITT) Standards
2400 bit/s	Synchronous, full duplex analogue or digital	V24, V28
4800 bit/s	Synchronous, full duplex analogue or digital	V24, V28
9600 bit/s	Synchronous, full duplex analogue or digital	V24, V28
48000 bit/s	Synchronous, full duplex analogue or digital	V35

The Data Link Level. The function of the data link layer is to take the basic transmission circuit provided by the physical layer and to transform it into a medium for reliable error-free transmission. Data link level protocols ensure that control information and user data are exchanged relatively free of error. This is achieved by breaking the input data into frames. Each frame contains within it a packet made up and delivered by the higher levels of the system. The structure of a typical frame is illustrated in Figure 1.15. The frame begins with a fixed 8-bit flag used for synchronization. Then follows an 8-bit field containing a destination address. This is followed by a control field which identifies the function and purpose of the frame and may contain a frame sequence number. The information packet then follows. After this, a frame check sequence (FCS) is appended. Typically, this will consist of 16 parity check bits generated by an error detection code. This is used to detect transmission errors. Finally, another 8-bit flag is appended to indicate the end of the frame. The data link level protocol specifies the frame structure, the elements of procedures (types of frame, frame sequence numbering, commands, and responses), and a description of procedures (for link set-up, link disconnect, and information transfer for resetting).

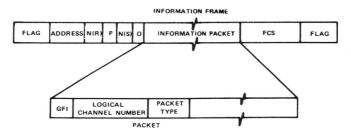

Figure 1.15   Packet and frame structures.

The Network Level. The network layer controls the operation of the communication subnetwork. It is sometimes called the communications subnet layer. The function of the network layer is to accept messages from the source computer, convert them into packets, and to control the transmission of those packets through the network to their destination.

The Transport Level. The transport level has the task of providing an efficient and reliable service between higher-level user processes. By contrast, the lower levels had the task of ensuring reliable communication between machines. As a specific case, many computers are multiprogrammed. In that case, multiple message streams will be flowing into the computer from the network, and vice-versa. The transport layer can provide the means for telling which message belongs to which connection.

The Session Level. An operation carried out by a computer user that involves more than one machine is usually called a session. For example, the term session may be used to describe the transfer of a file between two machines. The session level defines the procedures for setting up and management of a session.

The Presentation Level. The presentation level performs functions such as encryption (to provide security) and text compression (the efficient conversion of machine characters into a compressed binary form for efficient transmission).

The Application Level. The application layer is concerned with the network user programs and procedures. This layer separates the network users from the network designers.

1.3.5   Local Area Networks

A *local area network* can be described as a computer communications network that covers a limited geographical area. Compared to public data networks which operate over national and international boundaries, a local area network extends typically no more than a few kilometers. A local area network generally has an inexpensive communications medium based on cables or inexpensive digital radio equipment (17).

Public computer networks often do not offer the data transmission speeds required to allow modern computing, imaging, or storage devices to operate at full speed and are often not acceptable for the interactive transmission of program and data files in real time. Generally, a local area network is designed for operation at high data rates, typically from 100 kbit/s to 10 Mbit/s and higher.

Every node on the network can communicate with every other node using some form of "contention" procedure. Messages are "broadcast" over the communication network, with a destination address incorporated. All nodes have the capability of "listening in" to each message although only the intended receiver is expected to respond. Thus, a high level of privacy and security is not present unless cryptographic techniques are used.

The nature of most modern local area networks has been significantly affected by an important experimental research project conducted in the University of Hawaii in the 1970s. The project known as ALOHA used a packet-radio transmission system to connect terminals all over the Hawaiian islands to a computer and communications processor at the university, and from there to other public data networks including ARPANET.

The ALOHA packet-radio scheme was an innovative use of just two radio channels each operating on a single frequency (in a manner somewhat similar to a telephone party line). One of the channels is used for transmitting packets from the central computer to the remote terminals in a broadcasting mode. The destination address, which is contained in the packet header, permits reception at the addressed terminal. The other channel is shared by the remote terminals to transmit packets to the computer in a random access mode. Packets sent from different terminals may overlap in time (collide) and result in transmission errors. Retransmission is initiated when the terminal does not receive an acknowledgement from the computer within a time-out interval. To avoid repeated collisions, the delay before packet retransmission is randomized in each terminal.

This ALOHA concept was shown to be suitable for connecting a large number of terminals to a computer when the terminal output is bursty, that is, when the terminal is used to transmit short messages rather infrequently. The random access procedure just described is known as "pure ALOHA." In an alternative scheme, known as "slotted ALOHA," a sequence of time slots is established and packet transmissions are synchronized to fall within one of these time slots. This can increase the maximum throughput to twice as much as that for pure ALOHA.

Numerous refinements to the basic ALOHA scheme have been developed for local area networks, the most significant one being the *Ethernet* scheme developed at the Xerox Reseach Center. (Ether was historically thought to be a universal medium capable of transmitting electromagnetic waves. Xerox decided to build its "ether" out of coaxial cable.) In the Ethernet scheme, a station can sense when another station is sending data and stop transmitting before the end of its packet. A randomized delay function is incorporated so that each station waits a different amount of time before attempting a retransmission.

Generally, the algorithms such as Ethernet have to be designed to be as simple as possible to ensure minimum node costs and reliability. This is because each station on the network must manage its use of the network independently.

1.4   INTEGRATED SERVICES AND HIERARCHIES

In the design of a digital communication network, it is usually required to provide transmission systems suitable for digital signals from many different types of sources. These sources include telephones, telex equipment, voice-band modems, television equipment (video and audio), and baseband digital data signals associated with public data networks.

Since speech represents the major traffic in most public communication networks, it is usually desirable that bit rates associated with all other sources be integer multiples or submultiples of the bit rates associated with telephone transmission. We next consider some of these hierarchical relationships.

1.4.1 PCM Hierarchies

Pulse code modulation (PCM) systems for time-division-multiplex (TDM) telephone transmission were discussed in Section 1.1.3. It was noted that many countries have adopted the 30-channel PCM hierarchy which is based on each voice channel being encoded into 64 kbit/s and time-division multiplexed in groups of 30 voice channels. Two additional 64 kbit/s channels are used for signalling and synchronization. The result of encoding the 30 voice channels and two additional channels is referred to as a *primary multiplex* or *first-order multiplex*. For the 30-channel system, the first-order multiplex (30-channel) bit rate is 2048 kbit/s.

When several first-order multiplex signals are to be time-division multiplexed together to form a higher bit-rate stream, the arrangement of groupings must satisfy a standard hierarchical structure. This is illustrated in Figure 1.16. Note that when four 2048 kbit/s primary multiplex streams are multiplexed, the resultant second-order multiplex bit rate is 8448 kbit/s. This is greater than 4×2048 kbit/s because redundant bits equivalent to an additional time slot are added to each of the 30-channel 2048 kbit/s tributaries for framing purposes. More details are given in Chapter 2, Volume 2.

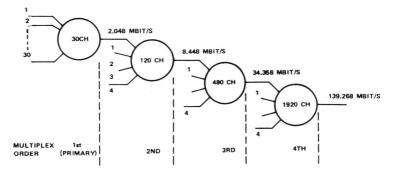

Figure 1.16 30-channel PCM hierarchy.

For convenience, the bit rates for the various multiplex levels shown in Figure 1.16 are often loosely referred to as 2 Mbit/s, 8 Mbit/s, 34 Mbit/s, and 140 Mbit/s, respectively. Many national digital radio and fiber optic transmission systems are being designed for bit rates of 140 Mbit/s.

In the U.S., U.K., and Japan, a 24-channel primary multiplex scheme is used. The bit rates for higher-order multiplexing hierarchies are given in Figure 1.17. Note that the schemes are similar except that higher-order multiplexing bit rates used in Japan differ from those currently adopted in the U.S. and U.K.

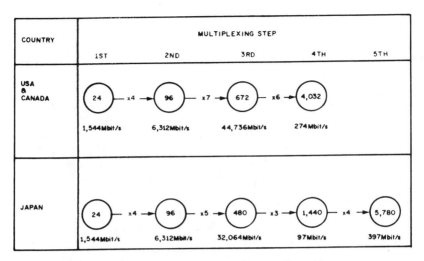

Figure 1.17 24-channel PCM digital hierarchies.

1.4.2 Bit Rates for Other Services

For the types of PCM systems described above, the standard bit rate per telephone channel is 64 kbit/s. Other analogue signals are also transmitted in a digital network after analogue-to-digital conversion (encoding). For convenient integration of these signals into a digital telecommunications network, it is desirable that the bit rates be integer multiples or submultiples of the 64 kbit/s single-telephone-channel rate.

Table 1.2 shows typical bit rates used for the encoding or multiplexing of various signals by commercially available equipment.

Specific examples from Table 1.2 include the following:

(1) Speech is encoded into 64 kbit/s using PCM or into 32 kbit/s using Adaptive Delta Modulation (ADM) or Adaptive Differential PCM (ADPCM).

(2) Telex and low rate asynchronous data up to 300 baud can be efficiently multiplexed by time division multiplex (TDM) Telex equipment into a 2400 bit/s stream.

(3) Data streams ranging from 1200 bit/s to 9600 bit/s are multiplexed by ''zero order'' multiplex equipment into 64 kbit/s streams.

(4) Sound program signals are encoded into integer multiples of 64 kbit/s. Typically five 15 kHz channels or ten 7 kHz channels are multiplexed into 2048 kbit/s.

Table 1.2
Bit rates for various signal sources.

DIGITAL TRANSMISSION RATES

SPEECH		64 kbit/s, 32 kbit/s
TELEX	50 baud	46 into 2400 bit/s
	75 baud	30 into 2400 bit/s
	200 baud	10 into 2400 bit/s
	300 baud	7 into 2400 bit/s
DATA	2400 bit/s	20 into 64 kbit/s
	4800 bit/s	10 into 64 kbit/s
	9600 bit/s	5 into 64 kbit/s
AUDIO PROGRAM		
	15 kHz	5 into 2.048 Mbit/s (6×64k ea.)
	7 kHz	10 into 2.048 Mbit/s (3×64k ea.)
PAL TV Currently		68 Mbit/s
	Future	34 Mbit/s

(5) Television signals can be encoded into 64 Mbit/s streams. It is anticipated that within the next few years this will be reduced to one 34 Mbit/s by the use of more sophisticated encoding schemes. For teleconferencing, much lower bit rates can be tolerated. For example, one system used in the U.S. operates at 3.152 Mbit/s.

Chapters 5 and 6, Volume 2 describe in detail the services to be provided in an ISDN environment.

1.5 PROBLEMS

1.1 Explain what is meant by the following functions carried out in a PCM system: Sampling, quantizing, companding, and multiplexing.

1.2 For a 24-channel and 30-channel PCM system, respectively, determine and compare the percentage of transmitted bits that represent overhead bits, that is, other than coded speech elements.

1.3 Illustrate how a single voice channel signal $x_T(t)$ is transmitted in a 30-channel PCM system by considering the segment of $x_T(t)$ given in Figure 1.18 and sketching the output from the sample-and-hold, 16-level uniform quantizer and multiplexer, respectively.

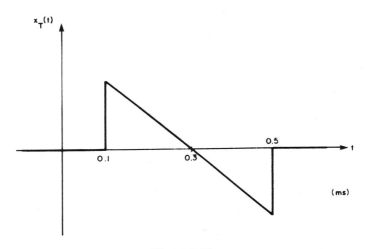

Figure 1.18

1.4 If a 5 MHz colour television signal is sampled at a rate of $f_s = 10$ MHz and quantized into 512 levels in a PCM system, find the transmission bit rate.

1.5 A certain telephone call in an analogue communication network passes through two local exchanges, four trunk exchanges, and one further local exchange in tandem. Four-wire FDM systems are used between the trunk exchanges.

    (i)   Determine the minimum number of modulation-demodulation steps involved.

    (ii) In a network, what are the advantages in minimizing the number of modulation-demodulation steps?

    (iii) If the call were made in a fully digital network, how many modulation-demodulation steps would be required?

1.6 (i)   Draw a suitable phasor design for a transmit signal space in an 8-PSK data transmission system.

    (ii) If a periodic binary data sequence . . . 101010 . . . is transmitted in the 8-PSK system, sketch the line waveform.

    (iii) Repeat (ii) for the case of a 16-QAM system with signal space given in Figure 1.11.

1.7 The input data sequence,

       0 1 1 1 0 0 1 0 0 1 1 1

is to be transmitted using

(i)   8-PSK

(ii) 16-QAM (CCITT Rec. V.29).

For each case, draw suitable transmit signal space phasor diagrams. Make Gray code assignments to each phasor.

Sketch the transmitter output waveforms for each case for the data sequence above.

1.8 Draw a block diagram to show how three 9600 bit/s baseband data signals and four 4800 bit/s sources could be multiplexed to form a simple composite 64 kbit/s signal.

1.9 (i)   If a 140 Mbit/s digital transmission system were used solely for 300 baud telex services, how many services could be accommodated?

(ii) If instead it were used for telephony using ADPCM, how many telephone signals could be accommodated?

1.6 REFERENCES

1. K. Iwasaki, "Digitalization of Japan's International Networks," *Proc. PTC'85*, pp. 22–29, Honolulu, January, 1985.
2. K. Feher, *Digital Communications : Microwave Applications*, Prentice-Hall, 1981.
3. M. Fournier, "What is a digital world?," *Proc. PTC'85*, pp. 4–10, Honolulu, January 1985.
4. T. Poussard, C. Beare, and Y. Falkovitz, "An introduction to the Australian Digital Data Network (DDN)," *Telecommunication Journal of Australia*, Vol. 31, No. 1, pp. 19–27, 1981.
5. W. H. Thurman, "Telecommunications Down Under," *IEEE Communications Magazine*, Vol. 20, No. 2, pp. 31–44, March 1982.
6. K. S. Shanmugan, *Digital and Analogue Communication Systems*, John Wiley, 1979.
7. J. E. Flood (Ed.), *Telecommunication Networks*, Peter Peregrinus Ltd., 1977.
8. J. G. Pearce, *Telecommunication Switching*, Plenum Press, 1981.
9. Technical Staff, Bell Telephone Laboratories, *Telecommunications Transmission Engineering*, Bell System for Technical Education, Winston-Salem, N.C., 1982.
10. S. Welch, *Signalling in Telecommunications Networks*, Peter Peregrinus Ltd., 1979.
11. H. Inose, *An Introduction to Digital Integrated Communication Systems*, Univ. of Tokyo Press, 1981.
12. O. Borgstrom, B. Andersson, A. Marlevi, J. Anas, and S. Braugenhardt, *Digital Telephony: An Introduction*, L. M. Ericsson Telephone Company, Sweden, 1977.
13. P. B. Bylanski, and D. G. W. Ingram, *Digital Transmission Systems*, Peter Peregrinus Ltd., 1980.

14. K. Feher, *Digital Communications: Microwave Applications*, Prentice-Hall, 1981.

15. N. A. Duc, "Data Transmission Developments and Public Data Networks," *Telecommunications Journal of Australia*, Vol. 29, No. 3, 194–205, 1979.

16. Telecom Australia, "Towards 2000," *Report of a Multidisciplinary Task Force*, Melbourne 1974.

17. A. S. Tanenbaum, *Computer Networks*, Prentice-Hall, 1981.

18. F. F. Kuo (Ed.), *Protocols and Techniques for Data Communication Networks*, Prentice-Hall, 1981.

19. M. Decina, "Managing ISDN through International Standards Activities," *IEEE Communications Magazine*, Vol. 20, No. 5, pp. 19–25, Sept, 1982.

Chapter 2

BASEBAND DIGITAL TRANSMISSION SIGNALS

2.1 INTRODUCTION

2.1.1 A Digital Transmission System

In this chapter, we begin our study of the various engineering problems associated with transmitting a serial binary signal from a source to a destination in a communications network. Figure 2.1 illustrates some of the signal processing functions that may be required in a single source-destination link. The binary signal may, for example, be generated by digitizing an analogue signal. Pulse code modulation (PCM) encoding of voice waveforms is one typical case. Digital encoding of a television video signal is another.

Before the binary information is transmitted to the destination, a number of signal processing functions may be performed on the source sequence as shown in Figure 2.1. These may include one or more of the following:

(1) *Source coding*—processing the raw source sequence to reduce any inherent redundancy or to encrypt the bit stream for security.

(2) *Channel coding*—providing for error control either by error detection and ARQ (automatic-repeat-request) or by forward error correction. Error control is primarily of importance in data links but not in telephony links.

(3) *Line coding*—scrambling, coding, and pulse shaping the binary signal to generate a line input signal with frequency spectrum suitable for transmission and with in-built capability for providing bit synchronization at the receiving end.

After transmission through the line, which may include one or more repeaters, the received signal is processed by the following:

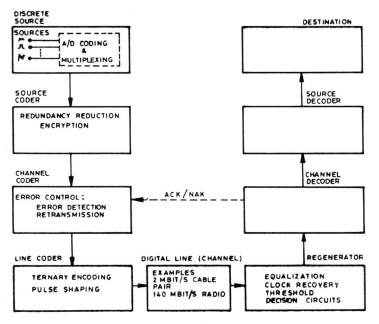

Figure 2.1   Digital transmission block diagram.

(1) *Regenerator*—amplifying, equalizing, sampling, and deciding which binary symbol was transmitted.

(2) *Channel decoder*—detecting errors and advising the transmitter or correcting errors.

(3) *Source decoder*—recovering the binary signal for delivery to the destination.

The block diagram in Figure 2.1 is primarily intended to clarify the functions carried out by different processing procedures in a digital communications link. In many practical links, there may also be one or more switching stages and the signal may be multiplexed with other signals.

Figure 2.1 is arranged to show layers or levels of signal processing. Note the similarity with the computer network protocol layers shown in Figure 1.14. In this and the next two chapters, we will concentrate on the bottom layer, shown in Figure 2.1, and referred to as the *baseband transmission* aspect of the system. In this chapter, we give particular attention to the problem of finding good forms of line transmission signals. In later chapters, we will examine other aspects.

2.1.2   Baseband Signals

A digital communications system must provide for efficient switching and transmission of digital signals. As discussed in the previous chapter, these digital signals may be associated with telephony (voice or signalling information), television (video and audio), computer data, or other services.

In this chapter, the point to point transmission of these digital signals is considered. Our primary concern is with line transmission over cable pairs or coaxial cable.

Consider the information signal at the line encoder input of Figure 2.1 to consist of a sequence of binary pulses, as illustrated in Figure 2.2. When such a signal is produced in a communication system, it is often referred to as a *baseband signal*. The concept of a baseband signal is commonly used but often loosely defined. We will refer to a baseband signal in a digital communication system as a signal for which the frequency spectrum extends to approximately zero frequency.

In the case of PCM telephony, such a baseband signal $v(t)$ may be generated in the PCM 30-channel primary multiplex equipment. The resultant baseband signal $v(t)$ is a binary signal at a bit rate of 2048 kbit/s. If four such 2048 kbit/s primary PCM signals are combined by time division multiplexing in a second-order multiplexer, an 8448 kbit/s baseband signal is produced.

When television video signals are encoded into digital form, the resultant baseband signal may be a binary signal at a bit rate of 68 Mbit/s. This signal can be considered to be of the same form as that shown in Figure 2.2. The spectrum of the multiplexed PCM signals will extend from zero to frequency values much higher than that of a single voice channel PCM binary signal.

The line coder shown in Figure 2.1 can be used to convert baseband binary signals into a more suitable form for transmission over line transmission media. As we will see later, the line coder can be designed to remove the dc component and attenuate low frequencies.

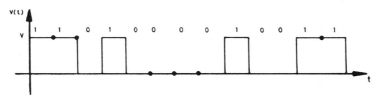

Figure 2.2 Input binary baseband signal.

2.2 BASEBAND LINE TRANSMISSION SYSTEMS

A block diagram of a digital transmission system used with cable pairs is shown in Figure 2.3(a). The system illustrated is typical of that used to transmit baseband PCM digits over balanced pair or quad cables.

The line transmission system incorporates a pair of line terminating units at each end of the link with repeaters spaced at appropriate intervals along the cable

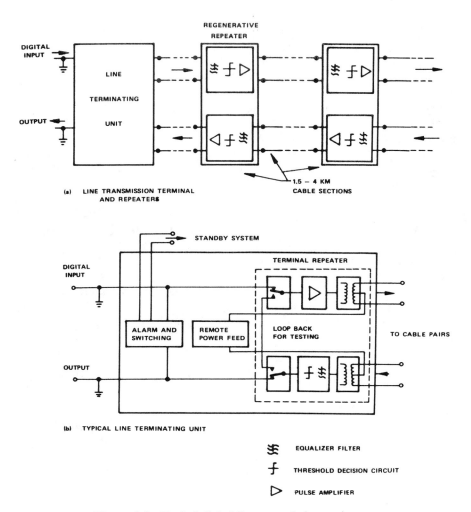

Figure 2.3 Typical digital line transmission equipment.

route. For cable pairs with conductor diameter in the range 0.4–1.4 mm, the repeater spacing could vary from 1.5 to 4km with attenuation of up to 40 dB per cable section (the attentuation being measured at a frequency of 1 MHz).

The functions of the line terminating unit are illustrated in Figure 2.3(b). The unit consists of transmitter and receiver sections which together are referred to as a terminal repeater. In addition, the line terminating unit contains ancillary facilities. These are the alarm and switching section, the remote power feed unit, and a facility to loop the receiver and transmitter sections for testing purposes.

The alarm and switching section monitors the operation of the line terminating unit. If it detects that the unit has failed, it can provide automatic switching of the digital input/output terminals to a standby unit. The remote power feed unit applies a dc voltage of up to 200 volts between the center taps of the transformers connected to the outgoing and incoming cable pairs, respectively. This can be used to provide power to repeaters located up to 120km from the line terminating unit.

The digital input to the line terminating unit passes through the transmitter section in which a pulse amplifier provides pulse amplification and pulse shaping. The signal is connected to the outgoing cable pair via a line transformer to ensure that the line is electrically balanced. This will minimize stray pickup (cross-talk) from adjacent pairs in the same cable.

After the signal has passed through a cable section, it is regenerated in a repeater. That is, the receiving section in the repeater reconstructs the original digital signal. This is achieved by an equalizer filter and a threshold decision circuit. The equalizer filter amplifies and filters the received signal in a way that compensates for the frequency-dependent attenuation that occurs in transmission. The threshold decision circuit is called a regenerator. It samples the received signal once each bit interval. Then it compares each sample against fixed threshold voltages in order to regenerate the original signal. The operation of the receiver filter and regenerator will be studied in detail in later sections.

After regeneration in the repeater, the signal is then passed through the transmitter section which consists of a pulse amplifier and line transformer as in the line terminating unit. This process is repeated after each cable section until the signal arrives at the line terminating unit at the receiving end. Here it is also equalized and regenerated in the same way as a regenerative repeater.

The resultant digital output is an error-free reproduction of the digital sequence at the input to the system provided that the noise and crosstalk encountered during transmission is not too large. This is ensured by limiting the maximum repeater section cable lengths. Note that regenerative repeaters reconstruct the digital signals at each repeater. They are not simply amplifying and equalizing repeaters as used in analogue transmission. In the following sections, we examine the digital transmission waveforms in detail.

2.3 ALGEBRAIC REPRESENTATION OF LINE SIGNALS

We now commence our examination of the types of signals suitable for baseband transmission. Some elementary forms of binary signal representation are first considered. The baseband waveform $v(t)$ which was illustrated in Figure 2.2 is known as a *unipolar NRZ* (nonreturn to zero) signal. Figures 2.4(a) and 2.4(b) show a *unipolar RZ* (return to zero) signal and a *polar RZ* signal, respectively.

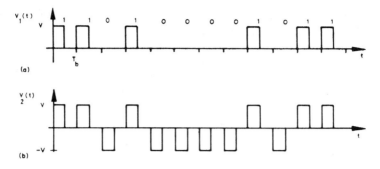

Figure 2.4   Unipolar RZ and polar RZ signals.

Note that the term bipolar is not used for the signal type illustrated in Figure 2.4(b). It is used to describe another signal format to be discussed in the next section. The signal waveforms in Figures 2.2 and 2.4 are not generally suitable for line transmission. To see why, we need to examine some of their properties. First we need to be able to express these types of signals in algebraic form.

Exercise 2.1

Write algebraic expressions for the random signals illustrated by the sample waveforms in Figures 2.2 and 2.4.

Solution. Consider first the unipolar NRZ waveform $v(t)$ in Figure 2.2. We first define a *basic pulse* shape $p(t)$ as follows

$$p(t) = \begin{cases} 1 & , \quad 0 < t \le T_b \\ 0 & , \quad \text{otherwise} . \end{cases} \tag{2.1}$$

Then we can write $v(t)$ in the form of a sum of pulses. That is

$$v(t) = \sum_{k=-\infty}^{\infty} a_k p(t - kT_b) \tag{2.2}$$

where the sequence

$$\ldots, a_{-1}, a_0, a_1, a_2, \ldots$$

represents a sequence of independent *random variables* taking the values 0 or V. In shorthand form, this is written

$$a_k \in (0, V) \text{ for } k = \ldots, -1, 0, 1, 2, \ldots \quad .$$

Now consider the unipolar RZ signal represented by $v_1(t)$ in Figure 2.4(a). If we define a new *basic pulse* $p_1(t)$ of width $T_b/2$ as follows:

$$p_1(t) = \begin{cases} 1 & , \quad 0 < t \le T_b/2 \\ 0 & , \quad \text{otherwise} \end{cases}$$

then we may write the unipolar RZ signal as

$$v_1(t) = \sum_{k=-\infty}^{\infty} a_k p_1(t - kT_b) \tag{2.3}$$

with $a_k \in (0, V)$.

Finally, to represent the polar RZ signal, we can use Equation 2.3 except that the random variables

$$\ldots, a_{-1}, a_0, a_1, a_2, \ldots$$

take the values $-V$ or V. That is, $a_k \in (-V, V)$.

An NRZ waveform is described as a 100 percent duty cycle pulse train. That is, a 1 is represented by a pulse of V volts for the whole duration of the symbol interval. On the other hand, an RZ waveform such as that described by Equation 2.3 is a 50 percent duty cycle pulse train.

2.4   ENCODING AND PULSE SHAPING

2.4.1   System Elements

We now consider in more detail the signal processing required to convert a "raw" binary signal (for example, unipolar NRZ) to a signal format suitable for transmission over a line system. Figure 2.5 illustrates the two-key signal processing steps required. The encoder converts the unipolar $(0, +V)$ binary signal into a line code signal such as at point A in Figure 2.5.

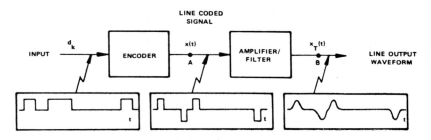

Figure 2.5   Line encoder and pulse shaping filter.

The transmit filter in Figure 2.5 then smooths out this line code signal to produce the line input waveform shown at point B. That is, it converts the theoretically infinite bandwidth signal at A into a bandlimited waveform for transmission.

2.4.2   Alternate-Mark-Inversion (Bipolar) Code

The line code signal represented at point A in Figure 2.5 is called an Alternate-Mark-Inversion (AMI) coded signal. It is also known as bipolar coding. It is used in some 2 Mbit/s and 1.544 Mbit/s PCM baseband transmission systems. The AMI signal is a ternary line code because each symbol is represented by one of three levels. The symbol 0 is represented by zero volts. The 1's are represented by alternate positive and negative pulses, referred to as "marks."

Now recall from Section 2.1.2 that a feature of the unipolar NRZ code that makes it unsuitable for transmission is that it contains significant dc and low frequency spectral components. Such a waveform would, therefore, suffer significant distortion during transmission over balanced lines with its associated coupling transformers. This effect is sometimes referred to as *dc wander* because it causes the peaks of the received pulses to vary slowly up and down depending on the density of marks in the waveform.

On the other hand, the AMI coded waveform does not have this disadvantage. Its dc component is zero since the $+V$ and $-V$ pulses occur equally often. We will also see in a later section that the AMI coding serves to reduce the low frequency spectral content.

In addition to the AMI code, there are many other line code waveforms that have been proposed. Some of these will be discussed later in this chapter. When designing a baseband digital transmission system, the line code is usually chosen so as to limit the low frequency components in the line waveform. That is, the line coder should generate a high-pass coded waveform without dc or low frequency components.

The pulse shaping filter is primarily required to shape and limit the high frequency spectral components of the line signal. This is done to minimize pulse distortion during transmission because of the high frequency roll-off effects in the transfer function of the line transmission line system. In Section 2.7 we will examine the line code properties in more detail. Pulse shaping filters will be discussed in Chapter 3.

2.5   LINE WAVEFORMS

It is instructive to examine the nature of the waveforms at the output of the line encoder and also after the pulse shaping filter shown in Figure 2.5.

Encoder output: A binary data sequence $\{d_k\}$ at the encoder input results in an encoder output which can be expressed

$$x(t) = \sum_{k=0}^{\infty} a_k p(t - kT_b) \qquad (2.4)$$

where for an AMI code, the a_k values can be $-V$, 0 or $+V$ volts. We have shown the index k starting at zero not at $-\infty$ as in Equation (2.2) to indicate that the first data bit occurs at $t = 0$.

Transmit filter output: Next we consider the pulse shaping filter to be such that when a pulse $p(t)$ is applied to the input, the filter produces a shaped pulse $p_T(t)$ for transmission on the line. The resultant line waveform $x_T(t)$ can, therefore, be represented as a sum of basic line pulses $p_T(t - kT_b)$, where $k = 0, 1, 2, \ldots$, each pulse being multiplied by the kth code symbol a_k. T_b is the bit interval. That is

$$x_T(t) = \sum_{k=0}^{\infty} a_k p_T(t - kT_b). \qquad (2.5)$$

For example, consider the use of an AMI code and a pulse shaping filter that produces a *basic line pulse* shape $p_T(t)$ given by

$$p_T(t) = \frac{\sin \pi r_b t}{\pi r_b t} \qquad (2.6)$$

where $r_b = 1/T_b$ is the bit rate.

It is convenient to write a shorthand form for this basic pulse using the *sinc function*, which is defined

$$\text{sinc}(A) = \frac{\sin \pi A}{\pi A}. \qquad (2.7)$$

The basic line pulse shape in Equation (2.6) can therefore be written

$$p_T(t) = \text{sinc}(r_b t).$$

This is illustrated in Figure 2.6 and is sometimes also called a *Nyquist pulse shape*. Nyquist shaping will be discussed further in Chapter 3.

Note that if $p(t)$ is a unit impulse, then the pulse shaping filter required to produce $p_T(t)$ will be an ideal lowpass filter of bandwidth $r_b/2$. Now the line waveform can be written

$$x_T(t) = \sum_{k=0}^{\infty} a_k \, \text{sinc}\{r_b(t - kT_b)\} \qquad (2.8)$$

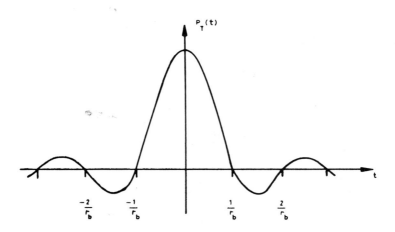

Figure 2.6 Sinc function basic line pulse.

where

$$a_k \in (-V, 0, +V).$$

We will see in a later section that the frequency spectrum of a sequence of sinc pulses of the form given in Equation (2.6) contains frequency spectrum components over the band up to $r_b/2$ (Hz). It contains no power above the frequency $B = r_b/2$ (Hz). That is, $p_T(t)$ is a low-pass signal which can be transmitted without distortion over a low-pass line which has a constant gain-frequency response from 0 to B (Hz). Note that in practice, $\text{sinc}(r_b t)$ pulse shapes can only be generated approximately. In principle, a $\text{sinc}(r_b t)$ pulse can only be generated by a filter with infinite delay.

Exercise 2.2

Consider the data sequence $\{d_k\}$ in Figure 2.5 to be (1 0 0 1 1 0 1 0 0). Assume that an AMI line code is used and the transmit filter produces the line waveform $x_T(t)$ of the form given by Equation (2.8). Compute values for $x_T(t)$ and plot as a function of time.

Note: It is suggested that this exercise be undertaken as a class exercise, each participant being expected to compute values of $x_T(t)$ for two points only. The points could be specified for example, as follows. Each person should compute $x_T(t)$ for values of $t = iT_b/5$ and $t = 2iT_b/5$, $i = 1,2,3 \dots$, where i is the particular participant's position on the class list. The final result should be assembled from the aggregate of calculated values.

Solution. Consider, for example, a participant whose name is listed in position $i = 13$ on the list. It is his/her task to compute $x_T(t)$ for values $t = 2.6T_b$ and $t = 5.2T_b$.

We consider here only the case $t = 2.6T_b$. We have nine input symbols so Equation (2.8) becomes

$$x_T(t) = \sum_{k=0}^{8} a_k \frac{\sin \pi r_b(t - kT_b)}{\pi r_b(t - kT_b)}$$

where the sequence $\{a_k\} = (a_0, a_1, a_2, \ldots a_8)$, will be

$$\{a_k\} = (+V, 0, 0, -V, V, 0, -V, 0, 0)$$

Note that we have arbitrarily assumed that the first 1 is represented by the symbol $+V$. Therefore, for $t = 2.6T_b$ we compute the sum

$$x_T(2.6T_b) = a_0 \frac{\sin \pi r_b(2.6T_b)}{\pi r_b(2.6T_b)} + a_1 \frac{\sin \pi r_b(2.6T_b - T_b)}{\pi r_b(2.6T_b - T_b)}$$

$$+ \ldots + a_8 \frac{\sin \pi r_b(2.6T_b - 8T_b)}{\pi r_b(2.6T_b - 8T_b)}$$

Now $r_b T_b = 1$ so we obtain

$$x_T(2.6T_b) = +V \frac{\sin 2.6\pi}{2.6\pi} + 0 + 0 - V \frac{\sin(-0.4\pi)}{-0.4\pi}$$

$$+ V \frac{\sin(-1.4\pi)}{(-1.4\pi)} + 0 - V \frac{\sin(-3.4\pi)}{(-3.4\pi)} + 0 + 0$$

$$= -0.7676V \text{ (volts)}$$

Table 2.1 lists a simple Basic program for these calculations, and Table 2.2 gives computed values of $x_T(t)$ for various integer multiples of $T_b/2$. Note that it is useful to compute a few $x_T(t)$ values for negative values of t as well as for positive t.

It may be instructive for readers to use such a program to compute $x_T(t)$ for other data sequences and for other line codes. For example, consider the unipolar code. Modify line 320 in Table 2.1.

The line coder output and the resultant line waveform $x_T(t)$ computed for the above AMI example are shown in Figure 2.7(a) and (b).

The line waveform $x_T(t)$ can be thought of as a sum of successively shifted Nyquist pulses (sinc pulses) each being multiplied by their associated a_k value $-V, 0$ or $+V$. These constituent sinc pulses are shown dotted in Figure 2.7(a).

Table 2.1
Basic program for computation of $x_T(t)$ values.

```
100   DIM A(8)
110   DEF FNSC(X)=(SIN(X))/X
120   FOR I=0 TO 8
130   READ A(I)
140   NEXT I
150   PI=3.14159
160   PRINT "t";"  X(t*Tb)"
170   PRINT "-";"  -------"
180   FOR A=0 TO 80 STEP 2
190   XT=0
200   T=A/10
210   FOR K=0 TO 8
220   Y=PI*(T-K)
230   IF ABS(Y)<.0001 GOTO 260
240   XT=XT+(A(K))*FNSC(Y)
250   GOTO 270
260   XT=XT+A(K)
270   NEXT K
280   PRINT USING "##.#";T;
290   PRINT USING "    ##.####";XT
300   NEXT A
310   STOP
320   DATA 1,0,0,-1,1,0,-1,0,0
```

The line waveform $x_T(t)$ in Figure 2.7(b) is the sum of these constituent sinc pulses.

An examination of Figure 2.7(b) indicates an advantage of using Nyquist pulse (sinc pulse) shapes. If we assume the waveform $x_T(t)$ in Figure 2.7 arrives undistorted at the regenerator input, then the regenerator can recover the best estimate of the input data sequence by proceeding as follows. First the regenerator samples the waveform $x_T(t)$ at times $t = 0, T_b, 2T_b, \ldots, 8T_b$. Then each sample value is compared against threshold voltage values set at $+0.5$ and -0.5 volt, respectively. If the sample exceeds $+0.5$ volt, then the regenerator decides that the transmitted a_k value was $+V$. If the sample value falls in the range between -0.5 and $+0.5$ volt, then the regenerator decision is 0. Finally, if the sample value is less than -0.5 volt, then the regenerator decides that $-V$ was the transmitted symbol. In the absence of noise or distortion on the channel, it can

Table 2.2
Computed line waveform values.

t/T_b	$X_T(t)$	t/T_b	$X_T(t)$
0.0	1.0000	4.2	1.2399
0.2	0.8517	4.4	1.2311
0.4	0.6104	4.6	0.9758
0.6	0.3454	4.8	0.5327
0.8	0.1263	5.0	0.0000
1.0	0.0000	5.2	-0.5108
1.2	-0.0241	5.4	-0.9030
1.4	0.0236	5.6	-1.1165
1.6	0.0844	5.8	-1.1385
1.8	0.0925	6.0	-1.0000
2.0	0.0000	6.2	-0.7618
2.2	-0.2035	6.4	-0.4943
2.4	-0.4835	6.6	-0.2582
2.6	-0.7676	6.8	-0.0903
2.8	-0.9661	7.0	0.0000
3.0	-1.0000	7.2	0.0269
3.2	-0.8269	7.4	0.0175
3.4	-0.4577	7.6	-0.0005
3.6	0.0420	7.8	-0.0083
3.8	0.5673	8.0	0.0000
4.0	1.0000		

be noted from Figure 2.7(b) that the sample values will be exactly the same as the input $\{a_k\}$ values. The data sequence can be recovered without error.

Each of the constituent Nyquist pulses that make up the waveform $x_T(t)$ extends over more than one bit interval of T_b (secs). That is, there is the potential for *intersymbol interference (ISI)* from one bit interval into others. This will usually be the case when a pulse shaping filter limits the high frequency components in the line signal. In doing so, time limited pulse shapes are smoothed and extended in time.

However, in the case of Nyquist pulses, even though each pulse will have a significant effect on line voltage values over adjacent bit intervals, there is nevertheless no ISI effect *at the ideal sampling instants*. This is because the sinc $(r_b t)$ pulse passes through zero at all time values iT_b units $(i = \pm 1, \pm 2, \pm 3, \ldots)$ on either side of the pulse center. That is, the zero crossings of any one Nyquist

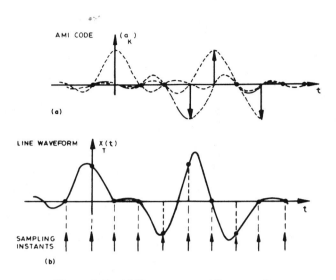

Figure 2.7 AMI sequence and line waveform.

pulse occurs at the sampling instants for all other input bits. We will see later that this is also the case for some other pulse shapes.

It is important to realize that typical digital line waveforms, as illustrated by Figure 2.7(b), are generally smooth because they are deliberately band-limited. At first sight, the waveform does not bear much resemblance to the original data input signal. Digital line waveforms are generally analogue-type waveforms and are not digital or discrete. It is the function of the regenerator to recover the data sequence by appropriate sampling and processing of the received line waveform.

2.6 LINE CODE SELECTION

In selecting an appropriate line code for baseband transmission, the nature of the particular transmission line/repeater system must be considered. Also, the type of signal regenerator at the repeater or receiving terminal should be taken into account. The essential task is to minimize the regenerator bit error rate.

2.6.1 Desirable Code Characteristics

In general, the choice of a particular line code usually involves consideration of the following:

(1) the shaping of the frequency spectrum of the line code waveform, particularly attenuation of dc and low frequency components

(2) the relative complexity of the encoder and associated decoder circuits

(3) the ability of the code to provide timing information at the regenerator, particularly in the event of long strings of 0's or 1's in the input data.

(4) the ability of the code to provide inherent error detection and an indication of regenerator decision errors. This allows simple monitoring of the system bit-error rate.

(5) the ability of the decoder to avoid propagation of a string of decoder errors due to isolated regenerator errors at the decoder input.

(6) the efficiency of the code in respect of maximum transmission rate within a given bandwidth.

In the following sections we will examine several line codes in relation to these requirements.

2.6.2 AMI Code Properties

The AMI code performs well in respect of items (1), (2), (4), and (5), but not (3). We will examine the frequency spectral properties of AMI coded sequences (item (1)) shortly. Before doing so, we will first consider the other properties.

Circuits for encoding and decoding are illustrated in Figure 2.8. The AMI encoder in Figure 2.8(a) consists of a modulo-2 adder, a delay element and a difference amplifier. The modulo-2 adder can be implemented by an exclusive-OR circuit. Modulo-2 addition of two binary symbols is defined as follows:

$$0 + 0 = 0$$
$$0 + 1 = 1$$
$$1 + 0 = 1$$
$$1 + 1 = 0$$

The difference amplifier performs conventional subtraction. That is, its output is the difference between its two input voltages (using normal algebra). The delay element can be implemented by a shift register clocked at the bit rate.

A decoder for an AMI code is shown in Figure 2.8(b). It consists of two threshold detectors (comparators) and an OR gate. Assume that a noise-free AMI input signal to the circuit can take one of three levels $-V$, 0, and $+V$. Then for one of the comparators shown, the threshold level could be set to $-V/2$ volt and that of the other to $+V/2$ volt. (These threshold levels provide a margin against malfunction resulting from noise.) Then the two comparators perform as follows. The output of the comparator in the upper part of the circuit goes high if the input voltage exceeds $+V/2$ volt. Otherwise it goes low. The bottom comparator is arranged to perform an inversion function. Its output goes low if

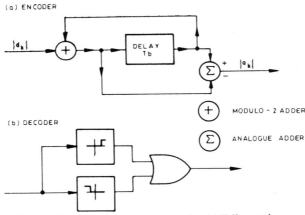

Figure 2.8 Encoder and decoder for AMI line code.

the input voltage exceeds $-V/2$ volt. Otherwise it is high. The operation of the circuit can be understood by considering a specific example.

Exercise 2.3

Let the $\{d_k\}$ sequence at the input to the encoder of Figure 2.8(a) be (10011). Show that the circuit produces an AMI output. If that output is fed to the decoder circuit of Figure 2.8(b), show that the circuit recovers the original data sequence.

Solution. We consider one by one the input $\{d_k\}$ sequence values. For convenience, let the output of the modulo-2 adder at time k be denoted c_k.

At the first symbol instant, one input to the modulo-2 adder will be $d_1 = 1$. The other input to the adder is assumed to be 0 (the delay element is assumed reset to zero when coding commences). Then the output will be $c_1 = 1$. As a result, the difference amplifier will obtain the output $a_1 = -1$.

At the next symbol interval, the input values to the modulo-2 adder will be $d_2 = 0$ and $c_1 = 1$ so its output will be $c_2 = 1$. Then the encoder output will be $a_2 = c_1 - c_2 = 0$.

In general, it can be seen that

$$a_k = c_{k-1} - c_k$$

where

$$c_k = d_k + c_{k-1} \ (\text{mod } 2).$$

Using these relationships, we can establish the sequences as follows

d_k	1	0	0	1	1
c_k	1	1	1	0	1
a_k	-1	0	0	1	-1

which provides the correct AMI coded output.

Next we examine the operation of the decoder circuit. The comparator thresholds are set to -0.5 and $+0.5$ volt, respectively. The first input value is -1 volt. As a result, the output of the top comparator will be low and that of the bottom will be high. Consequently, the output of the OR gate will be 1 as required. The next input is 0. Now the output of both comparators will be low so the OR gate output will be 0 as required. Likewise it is easy to see that if the input to the decoder circuit is $+1$ volt, the output will be 1, as required.

Next, let us examine the properties of the AMI code which enable the decoder to detect transmission errors. Such errors can occur at the regenerator as a result of transmission line noise. For example, consider the case where the transmitter sends the AMI symbol $-V$. At the regenerator the received line voltage is sampled. Let us assume the line to be lossless but noisy. At the sampling instant, a positive noise spike of magnitude greater than $V/2$ may occur. (Note that the system design should ensure that such occurrences are very infrequent.) However, if such a noise value occurred, the regenerator decoder circuit of Figure 2.8(b) would produce a 0 output instead of the binary 1 symbol that was intended.

Most error patterns are detectable at the decoder because of the inherent structure of the AMI code. For example, a single bit error would produce two successive pulses of the same polarity. This violates the alternate-mark rule. Detection of errors through code violations can provide a simple means of monitoring the bit-error rate performance of the transmission system. Note however that not all error patterns are detectable.

Exercise 2.4

The AMI coded sequence $(+1, -1, 0, +1)$ is to be transmitted over a digital line system. Because of noise and intersymbol interference, it is possible for the regenerator to make errors in up to two of the last three bits. Determine which error patterns are detectable as errors at the decoder and which are not. This exercise is left for the reader to complete.

At the decoder for the AMI code, error propagation cannot occur. That is, if the regenerator output contains an error, the associated decoder output bit may be in error. However, all subsequent output bits are decoded independently of any preceding bits.

Disadvantages of the AMI code

The main drawback of the AMI code is that there is a loss of regenerator synchronizing information for data streams with low mark density. If long strings of zeroes occur, there will be no transitions for synchronization in the clock recovery circuit at the regenerator. Note that if AMI line coding were used for a PCM system, then when an input channel is idle, the all-zero code word (00000000) could occur repetitively.

Another less significant disadvantage of the AMI code is its inherent low transmission efficiency. It is a three-level (ternary) code transmitting one bit per symbol. In principle, a three-level code should be capable of transmitting $\log_2 3 = 1.6$ bits per symbol. The AMI code is, therefore, approximately 60 percent less efficient than a binary code.

2.6.3 Manchester Code (twinned binary, split phase)

Another code sometimes used for digital transmission is the Manchester code, which is also called the twinned binary or split phase code. This code guarantees signal voltage transitions at least once each bit interval. It is, therefore, superior to AMI from the point of view of clock recovery. Figures 2.9(a) and (b) illustrate the coded sequence representations. Figure 2.9(c) illustrates the Manchester code for a sample data sequence. The Manchester code is sometimes used in local-

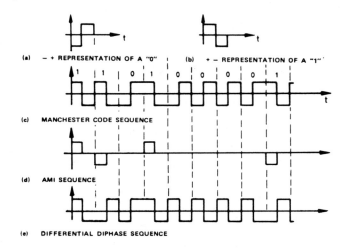

Figure 2.9 Illustration of Manchester codes.

area computer network systems such as one known as Ethernet (see Chapter 7). Figure 2.9(d) shows the equivalent AMI sequence for comparison.

The main disadvantage of the Manchester code is that it is less efficient in terms of bandwidth required for a given bit rate. This is indicated by the relatively high number of transitions in Figure 2.9(c) compared with 2.9(d). We will see later that a Manchester coded sequence occupies twice the bandwidth compared with an AMI coded sequence.

2.6.4   Differential Diphase Code

Another line code with provisions for easy clock recovery is the Differential Diphase (DD) Code. Actually, this code is sometimes referred to collectively with that in Figure 2.9(c) as making up the Manchester code group. In the DD code, each data bit is encoded with a transition in the center of the bit interval. In addition, a 0 is encoded with a transition at the beginning of each bit interval, while a 1 results in no such transition. Figure 2.9(e) illustrates this code. The DD code is used for digital transmission over balanced subscriber cable pairs (local loops) at bit rates of the order of 64 kbit/s.

Like the Manchester Code of Figure 2.9(c), the Differential Diphase code requires much larger bandwidth than the AMI code for the same bit rate. However, it does simplify the clock recovery process at the regenerator. This is the main advantage of using any one of the Manchester code group.

We will shortly examine other line codes that can avoid loss of synchronization because of low mark density but yet are not as spectrally inefficient, as are the Manchester or Differential Diphase codes. Firstly, however, we will review methods for determining the frequency spectral components associated with line codes.

2.7   METHODS FOR CALCULATING FREQUENCY SPECTRA

In many applications, the spectrum of the transmitted line signal must be accurately known. In this section we review the fundamental procedures for obtaining the spectrum. Then we examine the effect of line coding and pulse shaping, respectively, on the signal spectrum.

Different procedures are needed for finding the spectrum of a given signal depending on whether the signal is periodic, aperiodic, or random. The line codes are random waveforms in that it is not possible to predict what future data bits will be from past bits. However, it is worth reviewing the analytical procedures used for finding the spectra of each of the three types of signal.

2.7.1   Spectra of Periodic Signals

A periodic voltage signal $v(t)$ with period T_0 can be represented by a Fourier Series of the form

$$v(t) = \sum_{n=-\infty}^{\infty} c_n \cdot e^{j2\pi n f_0 t} \qquad (2.9)$$

where $f_0 = 1/T_0$ is the fundamental frequency and the Fourier coefficients c_n are the spectral values at the harmonic frequency $n f_0$.

Note that the exponential term $e^{j2\pi n f_0 t}$ represents a unit amplitude phasor rotating at $2\pi n f_0$ radians per second. Using the relationship

$$e^{jx} = \cos x + j \sin x$$

we can see that each $e^{j2\pi n f_0 t}$ term can also be envisaged as the combination of a cosine and sine function. The spectra values c_n are found using

$$c_n = \frac{1}{T_0} \int_0^{T_0} v(t) \, e^{-j2\pi n f_0 t} \, dt. \qquad (2.10)$$

The frequency spectrum of the periodic signal is of course a plot of the Fourier coefficients c_n (normally amplitude and phase plots). The amplitude values have the units of voltage.

Exercise 2.5

Determine and sketch the frequency spectrum for the periodic unipolar NRZ signal $v(t)$ shown in Figure 2.10 which represents the periodic data sequence ($\ldots 100010001000 \ldots$).

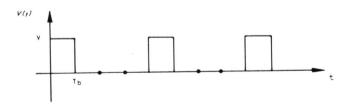

Figure 2.10   Periodic data sequence.

Solution. The spectral amplitudes c_n are obtained from Equation (2.10) as

$$c_n = \frac{1}{T_0} \int_0^{T_b} V \, e^{-j2\pi n f_0 t} \, dt$$

$$= \frac{V}{-j2\pi n f_0 T_0} \, e^{-j2\pi n f_0 t} \Bigg|_{t=0}^{t=T_b}$$

$$= \frac{V}{-j2\pi n} \cdot [e^{-j2\pi n/4} - 1] \,.$$

Now it is convenient to express the complex c_n values in terms of their magnitudes and phase using

$$\sin A = \frac{1}{2j} (e^{jA} - e^{-jA})$$

and so we rearrange the expression for c_n as follows

$$c_n = \frac{V}{\pi n} e^{-j\pi n/4} \cdot \frac{1}{2j} [e^{j\pi n/4} - e^{-j\pi n/4}]$$

$$= \frac{V}{\pi n} e^{-j\pi n/4} \sin(\pi n/4)$$

$$= \frac{V}{4} e^{-j\pi n/8} \frac{\sin(\pi n/4)}{(\pi n/4)}$$

and so obtain

$$c_n = \left\{ \frac{V}{4} \text{ sinc } \frac{n}{4} \right\} e^{-j\pi n/4}$$

The term inside the brackets is the modulus of c_n. It provides the voltage amplitude of the spectral lines and is shown plotted in Figure 2.11.

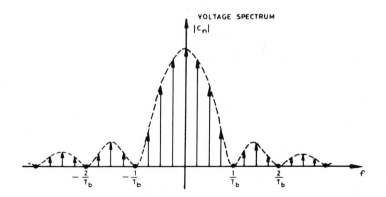

Figure 2.11 Amplitude spectrum of periodic data sequence.

We will find another plot very useful for spectral analysis of many signals, namely the *power spectral density* $S_X(f)$ of the signal $x(t)$. Parseval's theorem for periodic signals states that

$$\lim_{T \to \infty} \frac{1}{T} \int_{-T/2}^{T/2} |x(t)|^2 \, dt = \sum_{n=-\infty}^{\infty} |c_n|^2 \qquad (2.11)$$

The left-hand side of Equation (2.11) represents the normalized average power of the signal and $|c_n|^2$ gives the spectral distribution power.

Now we define the power spectral density function $S_X(f)$ to be a real, even, and nonnegative function of frequency that gives the distribution of power in the frequency domain. Then it follows that we can write

$$S_X(f) = \sum_{n=-\infty}^{\infty} |c_n|^2 \, \delta(f - nf_o) \qquad (2.12)$$

Note that $c_o\delta(t)$ represents an impulse at $f=0$ with magnitude (or weight) equal to c_o. This is the dc term. Also $c_1\delta(f-f_o)$ is the fundamental term at frequency $f_o \cdot c_2\delta(f-2f_o)$ is the second harmonic, and so on. That is, the power spectral density function $S_X(f)$ of a periodic signal is an impulsive function of frequency.

For the periodic data sequence of Figure 2.10, the power spectral density plot is shown in Figure 2.12. The power spectral values have units of Volt^2 and are the square of those in Figure 2.11. We note that for this case, most of the signal power is concentrated in the frequency band below $1/T_b$. That is, a low-pass transmission system with bandwidth $1/T_b$ would transmit most of the signal energy.

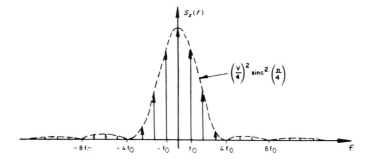

Figure 2.12   Power spectrum of periodic data sequence.

2.7.2   Spectra of Aperiodic Signals

If a signal $x(t)$ is not periodic but has finite energy, a convenient frequency domain representation may be obtained using the Fourier Transform

$$X(f) = \int_{-\infty}^{\infty} x(t) \, e^{-j2\pi f t} \, dt. \qquad (2.13)$$

This provides the spectrum of an aperiodic signal. The resulting frequency spectrum $X(f)$ is a *continuous* spectral function of frequency. The spectrum magnitude values $|X(f)|$ have the units of volts/Hz.

The functions $x(t)$ and $X(f)$ are said to constitute a *Fourier Transform pair*. Some useful Fourier transform pairs are given in Appendix 2.1.

Note that if the signal does not have finite energy, the integral in Equation (2.13) becomes infinitely large and so cannot be used to obtain the spectrum. Hence, the Fourier Transform method is primarily useful for those aperiodic signals that exist for a finite period of time and then decay away. An example is shown in Figure 2.13. It shows a signal $x(t)$ which is nonzero over an interval τ. Also shown is the magnitude plot of its Fourier Transform. The phase plot is usually of less interest and is not shown here.

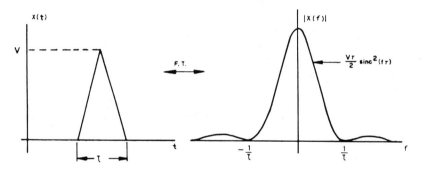

Figure 2.13　A finite energy aperiodic signal and its Fourier Transform.

For aperiodic signals, we define the energy spectral density $S_X(f)$ as

$$S_X(f) = |X(f)|^2 \tag{2.14}$$

which is a continuous function with the units of volts^2/Hz.

2.7.3　Spectra of Random Waveforms

Most signals transmitted in a communications system are random. That is, we cannot predict their future values even if we know their past values. The AMI coded waveform represented in Figure 2.7(b) is an example.

For random signals, it is not generally possible to compute the Fourier Transform because it does not exist. (The integral in Equation 2.13 "blows up.") Two alternative methods are available for estimating the frequency spectrum.

(1)   Auto-correlation Function Method

If we are given a signal voltage $x(t)$, we may be able to calculate the auto-correlation function $R_{xx}(\tau)$ of the signal. To do this, we can compute the ensemble average

$$R_{xx}(\tau) \;=\; E\{X(t)\,X(t+\tau)\} \tag{2.15}$$

for a collection $X(t)$ of sample waveforms (an ensemble) of the random signal process. We assume here that the signals are statistically "stationary." That is, their statistics are not affected by shifts in the time origin. This has been impled by writing $R_{xx}(\tau)$ as a function of the time lag τ only.

The process represented by Equation (2.15) is illustrated in Figure 2.14. This shows an "ensemble" of samples of a random AMI signal. We use the notation $X(t_1)$ to represent a random variable whose sample values are shown along the broken line at time t_1. Likewise, $X(t_1 + \tau)$ represents the random variable whose sample values are shown along the line at time $t_1 + \tau$. Note that we have used t rather than t_1 in Equation (2.15) since the choice of t_1 is arbitrary. The value of $R_{xx}(\tau)$ is independent of that variable for stationary signals. We compute Equation (2.15) using

$$R_{xx}(\tau) \;=\; \sum_{x_i}\sum_{x_j} X(t)\,X(t+\tau)P\{X(t)=x_i,\; X(t+\tau)=x_j\}.$$

Alternatively, it may be convenient for some random signals, to compute $R_{xx}(\tau)$ using a time average

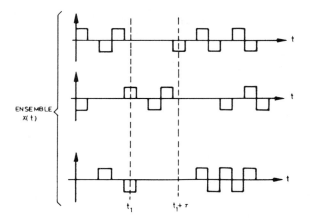

Figure 2.14   Calculation of $R_{xx}(\tau)$ using ensemble averaging.

$$R_{xx}(\tau) = \lim_{T \to \infty} \frac{1}{T} \int_{T/2}^{T/2} x(t) \, x(t+\tau) \, dt. \tag{2.16}$$

For more details, the reader should refer to a communication theory or random signal text. See, for example, Shanmugan (3).

Once we have found $R_{xx}(\tau)$ for the random signal, we define its power spectral density $S_X(f)$ as the Fourier Transform of $R_{xx}(\tau)$. That is,

$$S_X(f) = \int_{-\infty}^{\infty} R_{xx}(\tau) \, e^{-j2\pi ft} \, dt. \tag{2.17}$$

Exercise 2.6

Consider the periodic unipolar data sequence $x(t)$ represented in Figure 2.15. Find its autocorrelation function $R_{xx}(\tau)$. We observe that $x(t)$ is not a random signal, but we can still apply Equation (2.16).

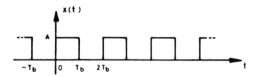

Figure 2.15 Periodic data sequence.

Solution. We illustrate the calculation of $R_{xx}(\tau)$ for one value of shift, namely $\tau = T_b/2$. Figure 2.16(a) shows the shifted version of $x(t)$, namely $x(t+T_b/2)$. Figure 2.16(b) shows the product function $x(t) \, x(t+T_b/2)$.

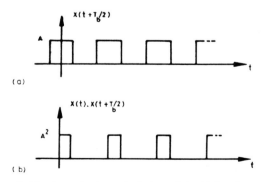

Figure 2.16 Steps in the calculation of $R_{xx}(\tau)$ by time averaging.

Now we obtain the value $R_{xx}(\tau)$ at $t = T_b/2$ by using Equation 2.16. This is easily done by finding the area under one period of Figure 2.16(b) with the result that

$$R_{xx}(T_b/2) = A^2/4.$$

Similarly, values for $R_{xx}(\tau)$ for other values of shift τ can be found. The complete result is shown in Figure 2.17.

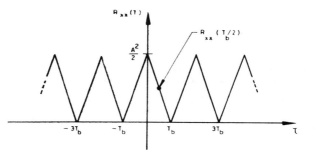

Figure 2.17   Autocorrelation function for a periodic data sequence.

Exercise 2.7

Generate a random binary waveform 20 bits long and estimate its autocorrelation function by time averaging.

Solution. Using 20 tosses of a coin, we can obtain a sample segment of random binary waveform, as illustrated in Figure 2.18(a). We use the same procedure to estimate $R_{xx}(\tau)$ as in Exercise 2.6. Figures 2.18(b) and (c) illustrate the steps in computing $R_{xx}(\tau)$ for $\tau = T_b/2$.

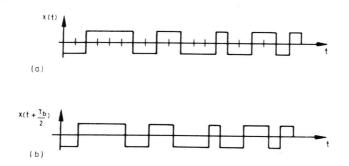

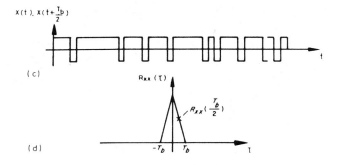

(c)

(d)

Figure 2.18 Estimation of $R_{xx}(\tau)$ for a random binary waveform.

Note that for an accurate estimate $R_{xx}(\tau)$ to be obtained, the length of the sample segment $x(t)$ must be much greater than the largest value of the shift τ that is to be used. This is because the product function $x(t)\,x(t+\tau)$ will only be defined for time values where sample functions $x(t)$ and $x(t+\tau)$ overlap.

Exercise 2.8

Consider the random binary waveform illustrated by the sample sequence in Figure 2.2. It can be shown that the autocorrelation function $R_{xx}(\tau)$ for this signal is given by

$$R_{xx}(\tau) = \begin{cases} \dfrac{V^2}{2}[1 - |\tau|/2T_b] & \text{for } |\tau| < T_b \\[2mm] \dfrac{V^2}{4} & \text{for } |\tau| \geq T_b \,. \end{cases} \qquad (2.18)$$

This is illustrated in Figure 2.19. Determine the power spectral density for this random binary signal.

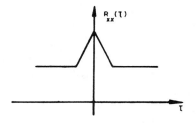

Figure 2.19 Autocorrelation function of random binary signal.

Solution. We need to obtain the Fourier Transform of $R_{xx}(\tau)$ given by Equation (2.17). One way to do this is to consider the function $R_{xx}(\tau)$ as consisting of two parts (illustrated in Figure 2.20). Then by superposition, the Fourier Transforms of each part can be added to obtain the final power spectrum as

$$S_x(f) = \frac{V^2}{4}\,\delta(f) + \frac{V^2}{2T_b}\,\text{sinc}^2(fT_b).\tag{2.19}$$

The first term in Equation (2.19) is the dc or average value of the signal $x(t)$.

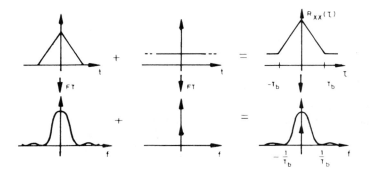

Figure 2.20   Calculation of power spectrum of random binary signal.

Often we are concerned to find the power spectral density of signals in discrete form, that is, defined in terms of discrete sample values. A discrete-time random signal can be described in terms of an ensemble of samples or snapshots as shown in Figure 2.21.

At a particular time, a random variable X_3, say, is defined as the set of samples at time $t = 3T_s$. A sequence of random variables

$$X_0, X_1, \ldots, X_j, \ldots$$

can be visualized, each being defined in terms of the values taken at times $0, T_s, 2T_s, \ldots$. The autocorrelation function $R_x(k)$ is defined as the expectation of the product of two random variables k samples apart. That is

$$R_x(k) = E\{X_j X_{j+k}\}.\tag{2.20}$$

As before, we assume the signal is stationary. Then as for the continuous signal case, the power spectral density $S_X(f)$ is defined as the Fourier Transform of $R_x(k)$ which may be written

$$S_X(f) = \sum_{k=-\infty}^{\infty} R_x(k)\, e^{-j2\pi fkT_s}\tag{2.21}$$

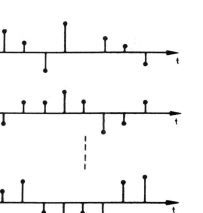

Figure 2.21 An ensemble for a discrete-time random signal.

In practice $R_x(k)$ is often estimated using the time average $R_N(k)$ of a finite-length segment of N samples where

$$R_N(k) = \frac{1}{N} \sum_{j=1}^{N-|k|} X_j \, X_{j+|k|}$$

and

$$R_x(k) = \lim_{N \to \infty} \{(R_N(k)\}.$$

(2) Fast Fourier Transform (FFT) Method

An alternative approach to estimating the power spectral density of a random waveform is by using the *Discrete Fourier Transform (DFT)* which is computed efficiently using the *Fast Fourier Transform (FFT)* algorithm. The signal must be in discrete form. That is, if it is as continuous signal, it must be sampled at a rate $(1/T_s)$ greater than the Nyquist rate. (The Nyquist rate is equal to twice the maximum frequency component in the signal.) Let the resultant time sequence $\{x(kT_s)\}$ consist of N samples spaced T_s seconds apart. Then the N-point discrete transform for this sample sequence is defined

$$X\frac{(nf_s)}{N} = \sum_{k=0}^{N-1} x(kT_s) \, e^{-j2\pi nk/N} \tag{2.22}$$

$$\text{for } n = 0, 1, \ldots, N-1$$

$$\text{where } f_s = 1/T_s.$$

The result of this transform $X \dfrac{(nf_s)}{N}$ is a set of discrete values at frequencies 0,

$1/NT_s$, $2/NT_s$, . . . , $(N-1)/N(T_s)$.

The power spectral density $S_X(f)$ of the original signal is then estimated from the sequence of $X \dfrac{(nf_s)}{N}$ values or by suitable averages of several DFT's of successive sample sequences of the signal. For more details, see for example, Shanmugan (3) or Miller (6).

2.8   POWER SPECTRAL DENSITY OF LINE CODES

2.8.1   Spectral Density of the Line Signal

We now consider methods for computing the frequency spectrum components associated with digital line waveforms. In doing this, the results of the previous section will apply.

For analytical convenience, it is useful to represent the encoder and pulse shaping process as shown in Figure 2.22. Note that the representation in Figure 2.22 differs from that in Figure 2.5 only in that the data sequence $\{d_k\}$ and the encoded sequence $\{a_k\}$ are represented by a series of delta functions of "weights" d_k and a_k. Then the filtered line waveform can still be represented by the series

$$x_T(t) \;=\; \sum_{k=-\infty}^{\infty} a_k\, p_T(t - kT_b)$$

as before.

The "basic pulse" shape $p_T(t)$ is the pulse obtained at the filter output when a positive impulse (delta function) is applied to its input. That is, $p_T(t)$ is the impulse response of the filter. We wish to find the power spectral density of the signal $x_T(t)$.

First we consider the sequence $\{a_k\}$ represented by a sequence of delta functions of weights . . . $a_{-1}, a_0, a_1, a_2, \ldots$. We assume that $E\{a_j\}=0$. Using Equation (2.20) we define the autocorrelation function $R_a(k)$ as

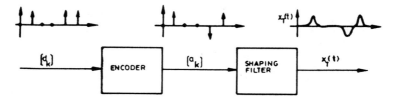

Figure 2.22   Encoder/filter model for spectral analysis.

$$R_a(k) = E\{a_j\, a_{j+k}\}, \tag{2.23}$$

for $k = \ldots -2, -1, 0, 1, 2, \ldots$.

The $R_a(k)$ values for k other than zero, will be a measure of the correlation between symbols in the sequence. For example, with AMI coding, the sign allocated to a mark may be positive or negative depending on the previous symbol allocated to a mark. That is, there may be some correlation between successive symbols. On the other hand, a purely random binary sequence will exhibit no correlation between adjacent symbols and $R_a(k)$ will be zero for values of k other than zero.

Then the time autocorrelation function for the impulse sequence $x(t)$ at the encoder output is also a sequence of impulses and is given by

$$R_{xx}(\tau) = \sum_{k=-\infty}^{\infty} R_a(k)\, \delta(\tau - kT_b).$$

The spectral density of $x(t)$ is given by the Fourier Transform of $R_{xx}(\tau)$.

To find this we first use the Fourier Transform of a single impulse (or delta) function located at $t = kT_b$. This is

$$F\{\delta(t - kT_b)\} = \int_{-\infty}^{\infty} \delta(t - kT_b) e^{-j2\pi ft}\, dt$$

$$= e^{-j2\pi fkT_b}.$$

Note that the last step follows from the "sifting property" of the delta function.

Then it follows that the spectral density of $x(t)$ is

$$S_x(f) = F\{R_{xx}(\tau)\}$$

$$= \int_{-\infty}^{\infty} \sum_{k=-\infty}^{\infty} R_a(k)\, \delta(\tau - kT_b)\, e^{-j2\pi f\tau}\, d\tau.$$

Exchanging the summation and integration operations, we obtain

$$S_x(f) = \frac{1}{T_b} \sum_{k=\infty}^{\infty} R_a(k)\, e^{-j2\pi fkT_b}.$$

Now since $R_a(k)$ must be an even function, we can write $S_x(f)$ in the form

$$S_x(f) = \frac{1}{T_b} \{R_a(0) + 2 \sum_{k=1}^{\infty} R_a(k)\, \cos 2\pi fkT_b\}. \tag{2.24}$$

This is the power spectral density of the signal $x(t)$ at the output of the line encoder. If the mean $E\{a_j\} = m$ where m is nonzero, then

$$S_x(f) = \frac{1}{T_b}\{R_a(0) - m^2 + 2 \sum_{k=1}^{\infty} |R_a(k) - m^2|\cos 2\pi f k T_b\} \qquad (2.25)$$

Unless otherwise stated, we assume that $E\{a_j\} = 0$.

$S_x(f)$ represents the power spectral density of the impulse sequence $d(t)$ at the input to the transmit filter. Now we can obtain the spectral density of the line signal $x_T(t)$ at the transmit filter output. Let $p_T(t)$ be the basic pulse shape in $x_T(t)$. It follows that $p_T(t)$ must be the impulse response of the transmit filter. Its Fourier Transform $P_T(f)$ is therefore the transfer function of the filter.

The filter output spectral density can be obtained from its input spectral density and its transfer function using

$$S_T(f) = |P_T(f)|^2 S_x(f). \qquad (2.26)$$

Finally, from Equations (2.24) and (2.26) we obtain the power spectral density of $x_T(t)$ as

$$S_T(f) = \frac{1}{T_b}|P_T(f)|^2\left\{R_a(0) + 2 \sum_{k=1}^{\infty} R_a(k) \cos 2\pi f k T_b\right\}. \qquad (2.27)$$

This important result can be used to compute the spectrum for any line coded signal used in baseband transmission.

2.8.2   Autocorrelation Function of Coded Sequences

To compute values for $R_a(k)$ defined by Equation (2.23), we consider coded symbols a_j and symbols a_{j+k} which occur k units later in the sequence. Then for each value of relative delay k, we can compute $R_a(k)$ providing we can determine the statistical interdependence of symbols separated k units in time. We need to know the set of values for the joint probabilities

$$P_{lm}(k) = P(a_j = l, a_{j+k} = m)$$

that is, the probability that a_j is some value l say, and also a_{j+k} is some other value m, where l and m are each one of the possible levels taken by the code. For example, if coded symbols can take the values $(-1, 0$ and $+1)$ then we find $R_a(k)$ using

$$E\{a_j a_{j+k}\} = \sum_{l=-1}^{1} \sum_{m=-1}^{1} lm\, P(a_j = l, a_{j+k} = m). \qquad (2.28)$$

Exercise 2.9

Compute $R_a(k)$ for the AMI sequence assuming that, at the encoder input, 0's and 1's occur with equal probability.

Solution. Consider first the autocorrelation for $k=0$, namely

$$R_a(0) = E\{a_j^2\}$$

$$= (-1)^2\, P(a_j=-1) + (0)^2\, P(a_j=0) + (1)^2\, P(a_j=1).$$

Since we assume 0's and 1's occur with equal probability at the encoder input, then for the coded output we have

$$P(a_j=-1) = 0.25 = P(a_j=1) \quad \text{and } P(a_j=0) = 0.5.$$

Hence, we obtain

$$R_a(0) = 0.5.$$

Now consider the computation of $R_a(1)$. To use Equation (2.28) we need the set of joint probabilities

$$P_{lm}(1) = P(a_j=l, a_{j+1}=m) \quad \text{for } l,m = -1,0,1.$$

That is, we have nine combinations of l and m values to consider. These probabilities are given in Table 2.3. The method of computation of each of these joint probabilities can be illustrated by the following cases:

(1) Computation of $P(a_j=-1, a_{j+1}=-1)$

The AMI code rules preclude a -1 symbol following a -1 so it is obvious that

$$P(a_j=-1, a_{j+1}=-1) = 0.$$

(2) Computation of $P(a_j=-1, a_{j+1}=0)$

To explain this calculation, for conciseness we define some "events" or possible outcomes as follows

$$\text{Let } A \text{ be the event that } a_j=-1$$

$$\text{Let } B \text{ be the event that } a_{j+1}=0$$

Then we can write

$$P(a_j=-1, a_{j+1}=0) = P(A,B).$$

Now we cannot compute $P(A,B)$ directly because event A (the probability that $a_j=-1$) depends on whether the previous mark was encoded as $+1$ or -1. To resolve this issue, we use the *conditional probability* $P(A,B|C)$ where we define C to be the event that the previous mark in the input sequence was coded as $+1$. Likewise we can refer to the conditional probability $P(A,B|\overline{C})$ where $\overline{C}$ is the event that the previous input mark was encoded as -1.

Table 2.3
Joint probabilities $P(a_j=1, a_{j+1}=m)$.

			a_j	
		-1	0	1
	-1	0	0.125	0.125
a_{j+1}	0	0.125	0.25	0.125
	1	0.125	0.125	0

With a little thought, it is clear that for the AMI code

$$P(A,B|C) = 1|4$$

and

$$P(A,B|\overline{C}) = 0.$$

Then to find $P(A,B)$ we use the well-known relationships from probability theory

$$P(A,B) = P(A,B,C) + P(A,B,\overline{C}).$$
$$= P(A,B\,|C)P(C) + P(A,B|\overline{C})P(\overline{C}).$$

Now $P(C)$ is the probability that the previous mark in the input sequence was coded as a $+1$, and since in a long sequence there is equal probability that input marks will be encoded $+1$ or -1, we obtain

$$P(C) = 1/2.$$

Hence, we obtain the result

$$P(a_j = -1, a_{j+1}=0) = P(A,B) = 1/8.$$

The computation of the remaining values in Table 2.3 follows in similar fashion. Table 2.3 can be described as the matrix of joint probabilities $P_{lm}(1)$.

As an aid, in the computation of $P_{lm}(1)$ values we can list all possible 2-bit sequences as shown in Table 2.4 and their probabilities conditioned on assuming the previous input mark a_p was encoded as a -1 and $+1$, respectively. No other $a_j a_{j+1}$ pairs are possible for the AMI code. That is, their probability is zero.

Then we can use Table 2.4 to find the joint probability values. For example, $P(a_j=0, a_{j+1}=0)$ can be found using the first and fifth lines of Table 2.4 giving

$$P(a_j=0, a_{j+1}=0) = 1/4 \times 1/2 + 1/4 \times 1/2 = 1/4.$$

Table 2.4
List of allowable $a_j\,a_{j+1}$ sequences and their conditional probabilities.

Coding a_p of previous mark	Allowable Sequences a_j	a_{j+1}	$P(a_j,a_{j+1}|a_p)$
−1	0	0	0.25
−1	0	1	0.25
−1	1	0	0.25
−1	1	−1	0.25
1	0	0	0.25
1	0	−1	0.25
1	−1	0	0.25
1	−1	1	0.25

Also, using line 2

$$P(a_j=0,a_{j+1}=1) = 1/4 \times 1/2 = 1/8$$

and so on. In this way we can find the values in Table 2.3

As a check on the values in Table 2.3, consider the sum of values in the first column. This must be equal to $P(a_j=-1)$ since we can write

$$P(a_j=-1) = P(a_j=-1, a_{j+1}=-1)+P(a_j=-1,a_{j+1}=0)$$

$$+P(a_j=-1,a_{j+1}=1)$$

$$= 0 + 0.125 + 0.125$$

$$= 0.25$$

as expected.

Likewise we can sum the other two columns to check that we obtain

$$P(a_j=0) = 0.5 \text{ and } P(a_j=1) = 0.25.$$

Also, we can sum each of the rows respectively to check that

$$P(a_{j+1}=-1) = 0.25, P(a_{j+1}=0) = 0.5 \text{ and } P(a_{j+1}=1) = 0.25.$$

Now we can complete our computation of the autocorrelation value $R_a(k)$ for $k=1$ using Equation (2.28) which becomes

$$R_a(1) = (-1)(1) \, P(a_j = -1, a_{j+1} = 1) + (1)(-1) \, P(a_j = 1, a_{j+1} = -1)$$

since the other seven terms are zero. Hence,

$$R_a(1) = -1/4.$$

Next we compute $R_a(k)$ for $k = 2$. As before, we must first determine the matrix of joint probabilities $P(a_j = l, a_{j+2} = m)$ for l and m taking values -1, 0 or 1. Table 2.5 lists all possible three-bit sequences and their associated conditional probabilities.

Table 2.5
List of allowable $a_j \, a_{j+1} \, a_{j+2}$ sequences.

Coding a_p of previous mark	Allowable sequences a_j	a_{j+1}	a_{j+2}	$P(a_j, a_{j+1}, a_{j+2} \mid a_p)$
-1	0	0	0	1/8
-1	0	0	1	1/8
-1	0	1	0	1/8
-1	0	1	-1	1/8
-1	1	0	-1	1/8
-1	1	0	0	1/8
-1	1	-1	0	1/8
-1	1	-1	1	1/8
1	0	0	0	1/8
1	0	0	-1	1/8
1	0	-1	0	1/8
1	0	-1	1	1/8
1	-1	0	1	1/8
1	-1	0	0	1/8
1	-1	1	0	1/8
1	-1	1	-1	1/8

Now we can compute the matrix of joint probabilities $P_{lm}(2)$. For example,

$$P(a_j = -1, \, a_{j+2} = -1) = 1/8 \times 1/2 = 1/16$$

from the last line in Table 2.5. Also,

$$P(a_j = 0, \, a_{j+2} = -1) = 1/8 \times 1/2 + 1/8 \times 1/2 = 1/8$$

using lines 2 and 12 in Table 2.5. In similar fashion we can obtain all other joint probabilities. The resultant matrix of $P_{lm}(2)$ values is given in Table 2.6.

Table 2.6
Joint probabilities $P(a_j = l, a_{j+2} = m)$

		a_j		
		-1	0	1
a_{j+2}	-1	1/16	1/8	1/16
	0	1/8	1/4	1/8
	1	1/16	1/8	1/16

Now we can compute $R_a(k)$ for $k=2$ using Equation (2.28) and we obtain

$$R_a(2) = (-1)(-1)(1/16) + 2(1)(-1)(1/16) + (1)(1)(1/16)$$

$$= 0.$$

Following a similar procedure it can be shown that

$$R_a(k) = 0 \text{ for all } k \geq 2.$$

Now, recall that autocorrelation functions are always even functions. That is $R_a(1) = R_a(-1)$, and so forth. Hence, we can plot $R_a(k)$ values against k. This is shown in Figure 2.23.

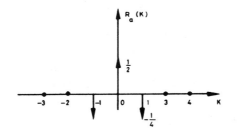

Figure 2.23 Autocorrelation function for AMI coded sequence.

2.8.3 Spectrum of AMI Coded Signal

We are now in a position to calculate the power spectral density at the output of the transmit filter in Figure 2.22 where an AMI code is used. The power spectrum $S_x(f)$ for the AMI coded *impulse sequence* is obtained from Equation (2.24) as

$$S_x(f) = 1/2 - 1/2 \cos 2\pi f T_b. \tag{2.29}$$

This is plotted in Figure 2.24.

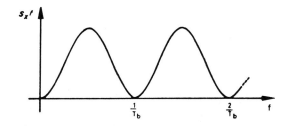

Figure 2.24   Power spectral density of the AMI impulse sequence.

Next we need to consider the effect of the transmit pulse shaping filter with impulse response $p_T(t)$ and transfer function $P_T(f)$. For simplicity, consider the case where $p_T(t)$ is a rectangular pulse as shown in Figure 2.25.

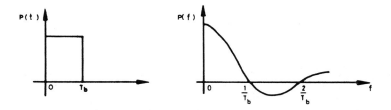

Figure 2.25   Impulse response and transfer function of the transmit filter.

Then the filter transfer function is (refer to Appendix 2.1):

$$P_T(f) = T_b \exp(-j\pi f T_b) \operatorname{sinc} f T_b.$$

In this case the line signal power spectral density $S_T(f)$ obtained from Equation (2.27) becomes

$$S_T(f) = T_b |\operatorname{sinc}^2 f T_b| [1/2 - 1/2 \cos 2\pi f T_b]. \tag{2.30}$$

The AMI line waveform $x_T(t)$ and its power spectral density $S_T(f)$ are shown in Figures 2.26(a) and (c). If instead of using the transmit filter with output basic pulse represented in Figure 2.25, we use an ideal low-pass filter of bandwidth $1/2T_b$ (Hz) then we have

$$P_T(f) = \begin{cases} T_b & |f| < 1/2T_b \\ 0 & \text{otherwise.} \end{cases}$$

In this case, the basic line pulse shape $p_T(t)$ will be

$$p_T(t) = \operatorname{sinc}(2t/T_b).$$

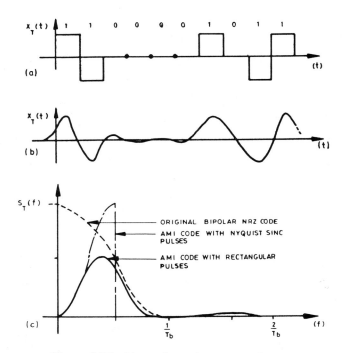

Figure 2.26 Line codes and power spectrum.

Figure 2.26(b) illustrates a typical line waveform for that case (see Exercise 2.2). Then the power spectral density is found from Equation (2.27) to be

$$S_T(f) = \begin{cases} T_b \, |1/2 - 1/2 \cos 2\pi f T_b| & \text{for} |f| < 1/2T_b \\ 0 & \text{otherwise.} \end{cases} \qquad (2.31)$$

This is also shown plotted in Figure 2.26(c).

2.8.4 Cyclostationary Signals

In the previous section we have obtained expressions for the power spectral density of line coded signals. The method involved the derivation of the auto-correlation function of the coded sequence. Then the associated power spectral density was obtained from the Fourier Transform of the signal's autocorrelation function. These statistical techniques provide us with a convenient method of obtaining the power spectrum. While they do provide useful solutions, the validity of the mathematical techniques requires some comment.

In Sections 2.8.1 to 2.8.3 we used autocorrelation functions of the form $E\{x(t_1) x(t_1 + \tau)\}$. Further, it was assumed that the autocorrelation functions for the digital signals are independent of the time origin t_1 and depend only on the relative shift τ. That is, the autocorrelation function could be represented by $R_{xx}(\tau)$, which has only one dependent variable τ. This was a necessary condition to enable the evaluation of the power specrum $S_x(f)$ of the signal via the Fourier Transform of $R_{xx}(\tau)$.

For the autocorrelation function of a random signal $x(t)$ to be independent of the time origin, the signal must belong to the class of stochastic processes known as wide-sense stationary (or weakly stationary) processes. These are defined as processes for which the ensemble mean $E\{x(t_1)\}$ and autocorrelation $E\{x(t_1)x(t_1 + \tau)\}$ are independent of the shift of time origin t_1. We have seen that baseband digital line signals can generally be written in the form

$$x_T(t) = \sum_k a_k p_T (t - kT_b).$$

Such signals by their nature contain periodicity in their structure. This is associated with the regular location of transition points once each symbol at intervals spaced T_b apart. If one of the transition points is maintained located at the time origin $t = 0$, then the process is known as a *cyclostationary process*. This is because the process exhibits a periodic time-varying ensemble mean, variance, and autocorrelation function.

On the other hand, if we "phase-randomize" the digital line signals, they become wide-sense stationary. This is a mathematical artifice in which the symbol transitions are maintained T_b apart but their locations are permitted to vary randomly with respect to the time origin. We could write such a signal mathematically as

$$x_T(t) = \sum_k a_k p_T(t - kT_b + \alpha)$$

where α is chosen to be a random variable uniformly distributed (chosen at random) between 0 and T_b.

Then we can find the signal's *average* autocorrelation function $R_{xx}(\tau)$ and *average* spectral density $S_x(f)$ by averaging over all values of the random "starting time" α.

Strictly speaking, the derivation of the power spectral density of the line coded signals (Equation (2.27)) should have taken into account the cyclostationary nature of the signals. As it turns out, the result would have been just the same. That is, for practical purposes the cyclostationary nature of the signals can be neglected.

However, there are some circumstances when this is not so. One such case occurs in the consideration of *the interference* as a result of crosstalk between digital systems, where decisions are made on periodic samples of the received signal. In this case, the relationship between sampling instant and the time-varying ensemble statistics must be taken into account. That is, the cyclostationary nature of the digital signals must be considered.

The study of cyclostationary processes is beyond the scope of this text. An excellent summary of their properties is given in Franks (7). The cyclostationarity of crosstalk interference from digital signals in multipair cable is discussed in Campbell (8).

2.9 OTHER TERNARY LINE CODES

We can now return to the consideration of ternary line codes which are of greatest significance for baseband digital line transmission. In Section 2.6, a list of desirable criteria was given for good codes. In summary, the factors to be considered are

(1) spectral distribution
(2) codec complexity
(3) bit timing content
(4) error detection
(5) error propagation
(6) transmission efficiency, and
(7) required equalizer bandwith (crosstalk noise susceptibility)

In Section 2.6 we considered the alternate-mark-inversion (AMI) code, the Manchester code, and the differential-diphase (DD) code.

Exercise 2.10

Summarize the advantages and disadvantages of the AMI code and the Manchester code class.

Solution.
AMI Code

Advantages:
(1) DC and low frequency components are attenuated
(2) Provides for error detection
(3) No error propagation occurs as a result of regenerator errors

Disadvantages:
(1) Loss of timing synchronization information with low mark density
(2) Transmission efficiency of one bit per symbol is 60 percent below maximum obtainable using a binary code.

Manchester Code and DD Code

Advantages:
(1) The coded sequence contains transitions in each bit
(2) DC and low frequency components are attenuated
(3) No error propagation occurs.

Disadvantages:
(1) The frequency spectrum (major lobe) extends to $2/T_b$ (twice as far as for AMI).
(2) No error detection capability inherent in the code structure.

Note that for the Manchester code, the power spectrum $S_x(f)$ is given by (see Problem 2.20):

$$S_x(f) = 1 - \cos \pi f T_b. \qquad (2.32)$$

The first null in this spectrum occurs at $f = 2/T_b$. We can compare this with the AMI code for which the first null is at $f = 1/T_b$.

In general, ternary line codes remain the most important class of codes for high-capacity baseband digital transmission. This is because their structure effects the attenuation of low frequency components. Line transmission systems usually exhibit a low frequency cutoff characteristic because of such things as input and output transformers at repeaters. The effect of attenuation of dc and low frequency components in a random binary pulse train is a varying zero symbol value. The term "dc or baseline wander" is used for this phenomena, which can lead to intersymbol interference. Ternary codes can minimize this effect.

Next we consider some important ternary codes which have been developed in the 1970s as alternatives to the AMI code.

2.9.1 High Density Bipolar (HDB*n*) Codes

A family of coding schemes known as HDB1, HDB2, HDB3, . . . , HDB*n*, . . . were proposed in the early 1970s. They are closely related to the AMI code. The most serious shortcoming of the AMI code is the lack of timing information for signals containing low mark density. The HDB*n* codes represent one of the most widely accepted solutions to this problem.

The basic concept is that when a run of more than n zeros occurs, the $(n+1)$th zero is replaced by a mark. This ensures timing information is always available no matter what data pattern is being transmitted. The most important of these codes is the HDB3 code.

HDB3 Code

The HDB3 Code is one of the most popular line codes. It is commonly specified for use at the primary multiplex output and higher order multiplexers up to the 34 Mbit/s multiplexer output. The HDB3 Code uses alternate-mark-inversion coding whenever possible but with modifications to avoid loss of timing with long strings of zeroes. The HDB3 Code is characterized by three rules as follows

Rule (1): If an input 0000 occurs, the fourth "0" is replaced by a substitution mark.

Comment: This guarantees bit timing information at least once every four input bits. However, it is necessary to distinguish the substitution mark from that for the input data sequences 0001.
This is accomplished by Rule (2).

Rule (2): Substitution marks are of the same polarity as the previous mark. That is, they violate the normal AMI rule.

Comment: Rules (1) and (2) are illustrated in Figure 2.27(a) in which the fourth 0 is represented by a positive pulse, violating the AMI rule. The next input 1 is represented by a negative pulse. One problem still remains. If the input consists of a very long sequence of 0's, we would have an output string of the form . . . $000V000V000V$ . . . where V represents the substitution (violation) marks. These must all be of the same polarity to satisfy Rule (2). The result would be that the coded sequence would contain low frequency energy down to zero

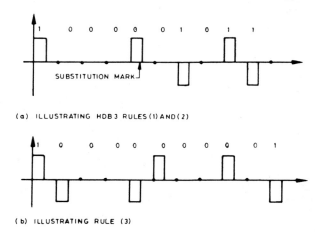

(a) ILLUSTRATING HDB3 RULES (1) AND (2)

(b) ILLUSTRATING RULE (3)

Figure 2.27   HDB3 line code.

frequency. To avoid this, we effectively alternate successive substitution marks (AMI violations) as specified by Rule (3).

> Rule (3): If the string 0000 occurs in the input binary stream, then either the two special sequences $000V$ or $M00V$ is transmitted, where M is a substitution mark obeying the AMI rule and V is a mark violating the AMI rule. $M00V$ is used if there has been an even number of ones since the last special sequence. Otherwise $000V$ is used.

This is illustrated in Figure 2.27(b). Eight successive 0's are represented by two $M00V$ sequences, where M represents an "obedient" substitution mark, and V represents a violation substituting mark. Note that, as a result of Rule (3), large strings of zeroes are encoded with alternating special sequences 1001 and $-100-1$. As a result, the dc and low frequency spectral values remain attenuated.

Exercise 2.11

For the following input data string, draw the associated HDB3 waveform. (Note that spaces are shown only for ease of reading sequence values.)

$$1011 \quad 1000 \quad 0101 \quad 0000 \quad 0000 \quad 1100 \quad 0001.$$

Assume that the basic line pulse $p_T(t)$ is a rectangular pulse of width $T_b/2$ as used in Figure 2.27.

Solution. The HDB3 coded signal is as shown in Figure 2.28. Note that the choice of the first $M00V$ or $000V$ sequence is arbitrary.

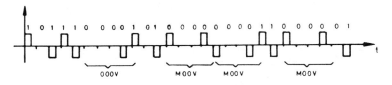

Figure 2.28 Example of an HDB3 signal.

The decoder for the HDB3 code must contain memory so as to check for two parameters. First, it must check for AMI violations. Secondly, it must check the number of zeros preceding this violation to determine if the last transmitted mark is also a substitution. As a result of this, it loses the attribute of instantaneous decoding, but only a small delay is involved.

Exercise 2.12

Decode the HDB3 line waveform shown in Figure 2.29.

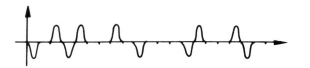

Figure 2.29 HDB3 waveform.

Solution. The decoded data sequence is

$$1011 \quad 0000 \quad 0100 \quad 0000 \quad 0010.$$

At first thought, it might appear that the error-detecting capability of the AMI code has been lost for an HDB3 code since an AMI violation no longer provides a unique indication of the occurrence of an error. However, careful examination of the effect of isolated single errors will show that they can be detected by the decoder. Each single error will either insert a spurious violation mark or will delete one of the deliberate violations.

The computation of the power spectral density $S_x(f)$ of an HDB3 line waveform using Equation (2.24) is tedious. Intuitively we might expect that $S_x(f)$ for the HDB3 code will not differ greatly from that for the AMI code. This is because the HDB3 and AMI codes are the same except when long strings of zeroes occur. The probability that in a particular four data bits, the sequence will be 0000 is only $(1/2)^4 = 1/16$ for equally likely 0's and 1's. Therefore, the probability that substitution marks occur in an HDB3 code will be small.

Figure 2.30 shows a plot of $S_x(f)$ for the HDB3 coding compared with the AMI code over the range $0 - 1/T_b$. As expected, the two spectra do not differ greatly. The average power in the HDB3 waveform is 10 percent higher than for the AMI since some of the 0 amplitude pulses in the AMI code are replaced by nonzero amplitudes in the HDB3 code. A lengthy but exact expression for $S_x(f)$ for the HDB3 code is given in Bylanski and Ingram (1).

2.9.2 MBNT Codes

The HDB3 code is commonly used in association with first, second, and third order PCM multiplex systems (that is, 2048 kbit/s, 8448 kbit/s, and 34368 kbit/s, respectively in the 30-channel multiplex systems). However, as pointed out previously, such a code sends binary bits using ternary symbols and is,

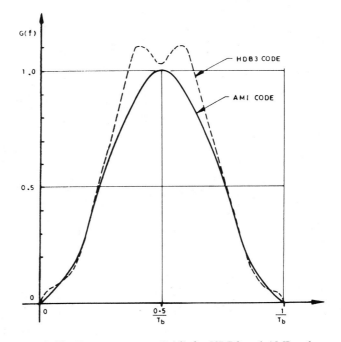

Figure 2.30 Power spectrum $S_x(f)$ for HDB3 and AMI codes.

therefore, less efficient than maximum as a result of the redundancy in the code. (The sign of the mark does not convey any information.) A three-level code should be capable of transmitting $\log_2 3 = 1.58$ bits per ternary symbol, so ternary codes increase the signalling rate by up to 58 percent over straight binary signalling.

For higher order multiplex systems, such as the 139,264 kbit/s fourth order system, the high bit rate is a severe constraint on transmission system design. For these cases, less redundant more complex line coding may be used. This may involve more efficient use of the ternary symbols. Alternatively, for certain transmission systems, particularly digital radio, M-ary signalling schemes may be considered. In these, each symbol may take on one of M possible levels $(M > 2)$. For example, for $M = 16$, each symbol can convey $\log_2 16 = 4$ bits of information. We will discuss M-ary signalling schemes at a later stage. Ternary codes remain the most important codes for high-speed baseband transmission.

We will now consider a class of more efficient ternary codes. They are known as the MBNT alphabetic codes. *Alphabetic* codes are those codes that map blocks of binary information bits into blocks of multilevel (in this case, ternary) symbols according to some state of the encoder. An *MBNT alphabetic* line code maps blocks of M binary information bits into blocks of N ternary channel symbols. Next we briefly review the most popular MBNT codes.

2.9.3 4B3T, 5B4T, 7B5T, 8B6T, and 10B7T Codes

The 4B3T code is one in which blocks of four binary bits are converted into blocks of three ternary symbols to reduce the symbol rate to 3/4 of the input bit rate. For a 140 Mbit/s system, this can reduce the symbol rate to 105 M sym/s. The code redundancy is reduced to 20 percent.

For four binary bits, there are $2^4 = 16$ possible combinations each of which is represented by three ternary symbols. As three ternary symbols provide $3^3 = 27$ possible combinations, there is a degree of freedom in choosing the code. The problem is to pair the binary and ternary groups so as to provide the desired characteristics of a line code set out in Section 2.6. Several schemes have been proposed for 4B3T codes. Bylanski and Ingram (1) give details of two such schemes.

Many other related ternary codes have been suggested for high-speed baseband transmission. For example, the 5B4T code converts five binary to four ternary symbols. Likewise, increased numbers of binary symbols can be converted in the 7B5T, 8B6T, and 10B7T codes.

Some codes are more efficient than others. For example, the 7B5T code ($2^7 = 128$ patterns chosen out of $3^5 = 243$ possible ternary patterns) is more efficient than the 5B4T code ($2^5 = 32$ patterns chosen out of $3^4 = 81$ possible patterns). However, for some transmission channels, the more efficient codes may do poorly at controlling dc wander because of low frequency cutoff effects. This may lead to significant ISI effects, which increase the bit-error rate. For a more detailed discussion, refer to Chang (4). Bylanski and Ingram (1) and Duc and Smith (5) provide an extensive bibliography of line transmission codes.

2.10 PROBLEMS

2.1 A general algebraic form for a baseband digital line waveform is given by Equation (2.2).

Sketch the waveform represented by Equation (2.2) for the data sequence defined for $k = -4, -3, \ldots, +4$. Consider the following cases. Note that $p(t) = 0$ except where defined below.

(1) $p(t) = 1, \ 0 < t < T_b/2$
and

$$a_{-4} = a_{-2} = a_0 = a_3 = 1,$$
$$a_{-3} = a_{-1} = a_1 = a_2 = a_4 = 0$$

(2) $p(t) = \sin^2(\pi t/T_b)$, $0 < t \le T_b$

and

$$a_{-4} = a_{-2} = a_0 = a_3 = 1,$$
$$a_{-3} = a_{-1} = a_1 = a_2 = a_4 = -1$$

2.2 A binary data sequence

1 0 0 1   0 1 0 1   1 0 0 0

is to be transmitted using the basic pulse shape $p_T(t) = 1$ for $0 < t < T_b/2$.

Sketch the line waveforms represented by Equation (2.5) for the following cases:

(1) AMI line waveform with a_{-1} assumed negative.

(2) AMI line waveform with a_{-1} assumed positive.

2.3 Repeat Problem 2.2, Part (1) for the case where $p_T(t) = \delta(t - T_b/2)$.

2.4 If the line waveform in Problem 2.3 is passed through the ideal low-pass receiver filter with voltage transfer function

$$H_R(f) = \begin{cases} 1 & \text{for } |f| < 1/2T_b \\ 0 & \text{otherwise} \end{cases}$$

sketch the filter output.

2.5 Repeat Problem 2.4 for the case where the filter bandwidth is

(1) $0.4/T_b$,   (2) $0.75/T_b$,   (3) $1/T_b$.

For each case, comment on whether a regenerator can obtain in each bit interval, a sample value which is free of intersymbol interference.

2.6 Repeat Problem 2.4 for the case where the filter is a first order low-pass system with

$$H_R(f) = 1/(1 + jf/f_0)$$

with 3 dB cut off frequency $f_0 = 1/T_b$ (H_z). Comment on the intersymbol interference.

2.7 Consider the AMI decoder in Figure 2.8. Devise additional circuit elements to provide an alarm output whenever a detectable error occurs during transmission. Describe how this might be used to estimate continuously the bit error rate on the transmission link. Is the estimate exact or likely to be a lower bound on the actual error rate?

2.8 For the AMI line code, compute the probability that a sequence of four line symbols contains one or more timing pulses (positive or negative symbol). Assume the data input binary symbols are equally likely.

2.9 Using the symbols + and − to represent a positive and negative pulse, respectively, consider the following line pulse sequence

$$+ - + - - + - + + - - + + - + - - +$$

Obtain the decoded binary sequence assuming the use of decoders suitable for

(1) Manchester coding
(2) Differential diphase

Assume that each bit is represented by a pair of symbols commencing at the first symbol given.

2.10 If the decoders in Problem 2.9 "slip" one symbol interval, that is, they assume that one of the bit intervals commences with the second bit rather than the first, determine the decoder output sequences in each case. Can the decoders detect that the slip has occurred?

2.11 It may be impossible on a given transmission medium to determine an absolute polarity. Repeat Problem 2.9 for the case where the line pulse sequence is inverted, that is, the polarities are reversed. Comment on which of the two coding schemes is more appropriate for this case.

2.12 Determine and sketch the frequency spectrum for the periodic unipolar signal similar to that in Figure 2.10, but with 1's represented by half-width positive pulses.

2.13 Determine and sketch the frequency spectrum of an AMI line waveform representing the periodic data sequence

$$\ldots 1 0 0 0 1 0 0 0 1 0 0 0 \ldots$$

if the basic pulse shape $p_T(t)$ is a half-width rectangular pulse of amplitude 1.

2.14 Repeat Problem 2.13 for the cases where the data sequence is

(1) $\ldots 1 0 1 0 1 0 1 0 \ldots$
(2) $\ldots 1 1 1 1 1 1 1 1 \ldots$

2.15 Determine and sketch the autocorrelation functions $R_{xx}(\tau)$ for each of the line waveforms described in Problems 2.12, 2.13, and 2.14.

2.16 Determine graphically and sketch the autocorrelation function $R_{xx}(\tau)$ for the AMI line waveform representing the "random" binary sequence illustrated in Figure 2.18(a). Assume the basic pulse shape $p_T(t)$ for the line waveform consists of full-width rectangular pulses (width T_b).

2.17 Repeat Problem 2.16 for the case where half-width rectangular pulses are used.

2.18 Use the Fourier Transform pairs of Appendix 2.1 to find good approximations to the frequency spectral amplitude distributions for the line codes of Problems 2.16 and 2.17, respectively.

2.19 Use Equation (2.22) to compute the Discrete Fourier Transform for the $\{x(kT_s)\}$ sequence

k	0	1	2	3	4	5	6	7
$x(kT_s)$	0	-1	-1	0	1	1	0	0

2.20 (1) For a Manchester coded sequence, compute the values for the auto-correlation function $R_a(k)$ defined by Equation (2.23).

(2) Hence, show that the spectral density of the Manchester coded signal is

$$S_x(f) = 1 - \cos\pi f T_b.$$

Sketch and compare this with $S_x(f)$ for the AMI sequence.

2.21 Find the spectral density $S_T(f)$ for an AMI line-coded waveform with half-width rectangular basic pulse shapes. Compare with other spectra shown in Figure 2.26.

2.22 Consider a unipolar random binary sequence such as the one labelled $\{d_k\}$ in Figure 2.22.

(1) Show that the autocorrelation function $R_a(k)$ takes the values 0.5 for $k=0$ and 0.25 for all other integer values of k.

(2) Hence, obtain the frequency spectrum $S_T(f)$ of a unipolar random waveform using half-width rectangular pulses such as illustrated in Figure 2.4.

Sketch $S_T(f)$ and show that it contains discrete frequency components at frequencies $1/T_b$, $3/T_b$, $5/T_b$, ...

(3) Show that if full-width rectangular pulses are used, there are no discrete frequency components.

2.23 Using the symbols $+$, 0, $-$ to represent a positive pulse, no pulse, and a negative pulse, respectively, determine the HDB3 line sequence for the binary data sequence

1 0 1 1 0 0 0 0 0 1 0 0 0 0 0 0 0 0 1 1

2.24 Decode the following HDB3 line sequence

$$+ 0 - + - 0\ 0\ 0 - + 0 - + 0 - + 0\ 0 + - 0\ 0 - 0\ 0 + 0 - +$$

where the symbols are used as described in Problem 2.23.

2.25 What is the average pulse density of HDB3 coding, that is, the probability that a given line symbol will be other than 0? Assume 0's and 1's are equally likely at the encoder input. Compare with the AMI code.

APPENDIX 2.1
TABLE OF FOURIER TRANSFORM PAIRS

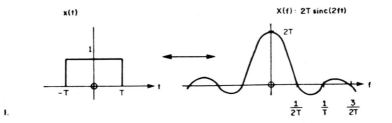

1.

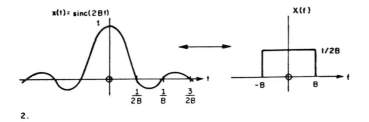

2.

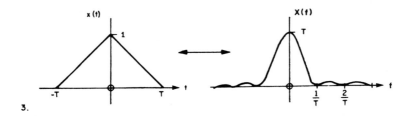

3.

APPENDIX 2.1
TABLE OF FOURIER TRANSFORM PAIRS (CONTINUED)

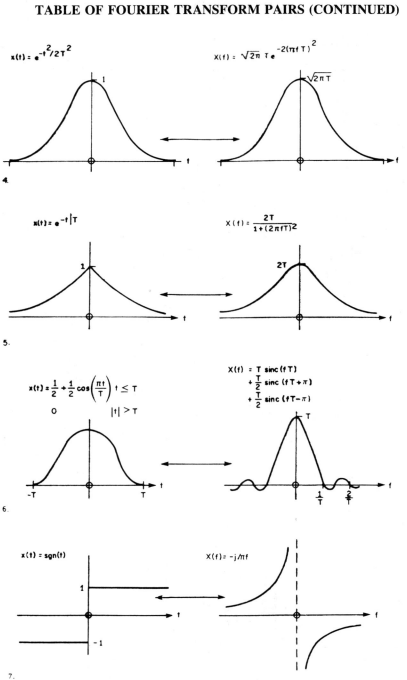

$x(t) = e^{-t^2/2T^2}$

$X(f) = \sqrt{2\pi} \; T e^{-2(\pi f T)^2}$

4.

$x(t) = e^{-t}|T$

$X(f) = \dfrac{2T}{1+(2\pi fT)^2}$

5.

$x(t) = \dfrac{1}{2} + \dfrac{1}{2}\cos\left(\dfrac{\pi t}{T}\right) \; t \leq T$

$0 \qquad |t| > T$

$X(f) = T \text{ sinc}(fT)$
$+ \dfrac{T}{2} \text{ sinc}(fT + \pi)$
$+ \dfrac{T}{2} \text{ sinc}(fT - \pi)$

6.

$x(t) = \text{sgn}(t)$

$X(f) = -j/\pi f$

7.

APPENDIX 2.1
TABLE OF FOURIER TRANSFORM PAIRS (CONTINUED)

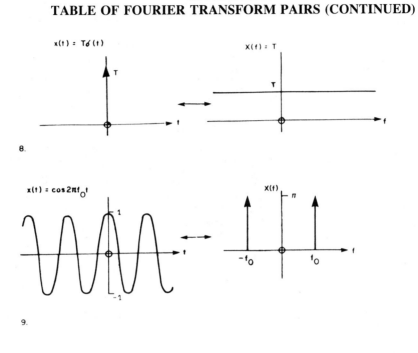

8.

9.

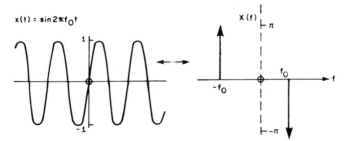

10.

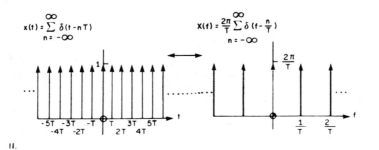

11.

2.11 REFERENCES

1. P. Bylanski and D. G. W. Ingram, *Digital Transmission Systems*, Peter Peregrinus Ltd., 1979.

2. A. Brinin and P. Frueh, *Transmission Systems Technology—PCM*, Telecom Australia Engineer Development Programme Publication, ED0027, 1980.

3. K. S. Shanmugan, *Digital and Analogue Communication Systems*, Wiley, 1979.

4. R. W. S. Chang, T. M. Jakubow, and A. Leon-Garcia, "Line Code Design for High-Capacity Baseband Digital Transmission Systems," *IEEE Trans. on Comms.*, Vol. COM-30, No. 7, pp. 1668–1678, July 1982.

5. N. Q. Duc and B. M. Smith, "Line Coding for Digital Data Transmission," *Australian Telecom Research*, Vol. 11, No. 2, 1979.

6. M. J. Miller, "Discrete Signals and Frequency Spectra," Chapter 5 of *Handbook of Measurement Science*, Vol. 1, Ed. P. H. Sydenham, Wiley, 1982.

7. L. E. Franks, *Signal Theory*, Englewood Cliffs, N.J.: Prentice-Hall, 1969.

8. J. C. Campbell, A. J. Gibbs, and B. M. Smith, "The cyclostationary nature of crosstalk interference from digital signals in multipair cable—Parts I and II," *IEEE Trans. on Comm.*, Vol. COM-31, No. 5, pp. 629–649, May 1983.

Chapter 3

INTERSYMBOL INTERFERENCE AND PULSE SHAPING

3.1 INTRODUCTION

In the previous chapter, we examined various line codes used in baseband digital transmission. An appropriate line code can ensure that bit timing information and error detection is available at the receiving end. Also, codes can be chosen that have frequency spectra in which dc and low frequency energy is attenuated so that intersymbol interference (ISI) resulting from baseline wander can be minimized.

We now turn to an examination of another source of errors in digital transmission systems. This is the problem of ISI resulting from distortion caused by channels with nonideal transfer functions at high frequencies. It is of primary concern in relation to maximizing transmission rates since in general, as the symbol rate is increased, so does the ISI because of high frequency pulse distortion.

Figure 3.1 shows the essential elements in a single repeater section of a baseband digital transmission system. At the transmitter an input binary data sequence $\{d_k\}$ is first encoded using a line code selected as discussed in the previous chapter. We will assume that a ternary code such as AMI or HDB3 is used. The resultant ternary sequence $\{a_k\}$ is then passed through a transmit filter which filters and amplifies the pulse train prior to transmission over the line or channel. The channel input waveform is denoted $x_T(t)$.

The transmission channel is characterized in Figure 3.1 as a linear system with additive Gaussian noise. The transfer function of the channel is denoted $H_C(f)$. The additive noise may represent thermal noise in the regenerator together with multiple sources of crosstalk from other channels. (If the crosstalk derives from a large number of independent sources, the central limit theorem of statistics allows us to conclude that the noise will be Gaussian no matter what the probability densities are of the individual sources.)

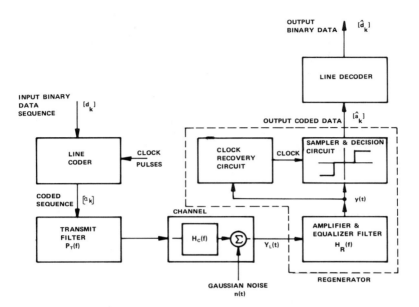

Figure 3.1   Baseband binary data transmission system.

The received signal $y_L(t)$ then passes into a regenerator where it is first amplified and equalized in an attempt to overcome distortion caused by the channel. The resultant waveform $y(t)$ is then sampled once each symbol interval. A threshold decision circuit compares each sample value against two thresholds in an attempt to regenerate the transmitted ternary sequence. The regenerated sequence is denoted $\{\hat{a}_k\}$. The sampling instants are determined by a clock or timing signal. This is generated from the waveform $y(t)$ by a clock recovery circuit. Finally, the regenerated ternary sequence is decoded to produce the output binary data sequence $\{\hat{d}_k\}$.

A set of typical waveforms at various points in the system is shown in Figure 3.2. Figure 3.2(a) shows the sequence produced by the line coder. An HDB3 code is assumed. As discussed in Chapter 2 of Volume 1, note that in the line coded sequence, binary 1's are shown encoded as a positive or negative pulse. It is convenient for analysis to assume that the pulses representing binary 1's at the transmit filter input are unit impulses as illustrated in the figure. Figure 3.2(b) shows a typical waveform obtained at the transmit filter output.

After transmission through the channel, the resultant waveform may be somewhat distorted. It may also contain additive noise. This waveform is passed through the regenerator amplifier and equalizer. Figure 3.2(c) illustrates the equalizer output waveform. This waveform is next sampled under the control

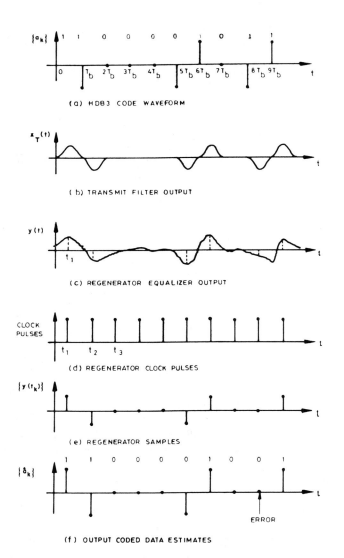

Figure 3.2 Typical waveforms in a baseband transmission system.

of a clock signal as shown in Figure 3.2(d). The resultant sample pulses are shown in Figure 3.2(e). Figure 3.2(f) shows the output coded data estimates obtained by the decision circuit on the basis of the sample pulses.

Note that in Figure 2.3, the ninth symbol regenerated in the sequence is in error. That is, a -1 value was transmitted. However, because of noise and/or distortion during transmission, the associated regenerator sample value was closer to a 0 symbol value than to a -1. The sample value in that case fell on the wrong side of the lower decision threshold. The regenerator decision circuit, therefore, decided that it represented a 0 and in doing so, made an error.

We will be particularly concerned in this chapter with the following:

(1) *Transmit filtering*—The ternary coded sequence $\{a_k\}$ is passed through a transmit filter with transfer function $P_T(f)$ and then into the transmission line. The main purpose of the transmit filter is to limit the high frequency spectral content of the signal to control the amount of pulse distortion in transmission.

(2) *Pulse dispersion*—The channel attenuation and group delay may cause pulse distortion. If the line waveform $x_T(t)$ were to consist of rectangular pulses, the attenuation and group delay effects of the line would cause the pulses to be spread out or dispersed. The transfer function of the transmit filter $P_T(f)$ and that of the regenerator equalizer filter $H_R(f)$ must be designed to control the amount of pulse dispersion that results from the nonideal transmission channel with transfer function $H_C(f)$.

(3) *Equalization and regeneration*—If the transfer function $H_C(f)$ of the channel was known completely at all times, it would be possible to design a system in which the pulse dispersion effects are zero. This is not possible in practice because of such factors as imperfect filter design, incomplete knowledge of the channel characteristics, and changes in those characteristics with time. In order to mitigate the residual distortion, it is usual to include in the system an adjustable equalizer filter to compensate for the distortion. This will minimize the number of regenerator errors.

(4) *Intersymbol interference (ISI)*—The effect of pulse dispersion may cause the pulse at the regenerator sampler input to be spread over several sample intervals. As a result, the sample value for one symbol is dependent not only on the transmitted symbol but also on symbols preceding and following it. This effect is called intersymbol interference (ISI). The ISI is data pattern dependent. That is, its magnitude and sign depends on the particular pattern of data that is transmitted in the vicinity of the sampling instant. Unless controlled by the transmit filter and receiver equalizer, the ISI can give rise to significant deterioration of regenerator output bit-error rate.

(5) *Noise and crosstalk*—The overall bit-error rate may also be affected by noise and crosstalk from other transmission lines. This can have two detrimental effects on regenerator decisions. Firstly, the recovered clock pulses may exhibit random timing jitter. Secondly, the sample values of the received waveform may differ so much from the transmitted pulse values that, as a result, regenerator decision errors are made.

In this chapter we examine each of these problems to see how low bit-error rates can be achieved in high bit-rate systems by appropriate design of transmit filters and receiver equalizers. As discussed in Chapter 2, we can write the line input waveform as

$$x_T(t) = \sum_{k=-\infty}^{\infty} a_k p_T(t - kT_b) \tag{3.1}$$

where $p_T(t)$ is the "basic pulse" associated with the transmit filter shaping. For convenience we will assume $p_T(t)$ is normalized such that $p_T(0) = 1$. The amplitude a_k depends on the data input d_k and the line code being used. For the HDB3 ternary code, a_k can take on values $-V$, 0 or $+V$ volts.

After passage through the transmission line and regenerator amplifier/equalizer, the received waveform can be written

$$y(t) = \sum_{k=-\infty}^{\infty} A_k p_r(t - t_d - kT_b) + n_o(t) \tag{3.2}$$

where $p_r(t - t_d)$ is the basic pulse output of the line and equalizer, t_d is a time delay and $A_k = K a_k$ with K a normalizing constant that yields $p_r(0) = 1$. The term $n_o(t)$ represents the effect of noise and crosstalk at the equalizer output.

The regenerator samples the waveform $y(t)$ at sampling instants

$$t_m = mT_b + t_d \quad , \, m = \ldots -1, 0, 1, 2, \ldots .$$

The sampled input to the decision circuit at sampling instant t_m is from Equation (3.2)

$$y(t_m) = \sum_{k=-\infty}^{\infty} A_k p_r(t_m - t_d - kT_b) + n_o(t_m)$$

which we can write

$$y(t_m) = \sum_{k=-\infty}^{\infty} A_k p_r((m - k)T_b) + n_o(t_m). \tag{3.3}$$

Ideally, at the sampling instants, only the mth input pulse ($k = m$) should contribute to $y(t_m)$. That is, since $p_r(0) = 1$, if the ISI resulting from other samples is neglected, the sample value would be given by

$$y_d(t_m) = A_m + n_o(t_m). \tag{3.4}$$

However, from Equation (3.3), the ISI contributed by other pulses is given by

$$y_I(t_m) = \sum_{\substack{k=-\infty \\ k \neq m}}^{\infty} A_k p_r((m - k)T_b). \tag{3.5}$$

Gathering these terms together, the value of the sample $y(t_m)$ at the regenerator decision circuit input is

$$y(t_m) = A_m + \sum_{\substack{k=-\infty \\ k \neq m}}^{\infty} A_k p_r((m-k)T_b) + n_o(t_m) \tag{3.6}$$

where the second term represents the residual ISI effect of all other bits on the mth bit being decoded. Ideally, the transmit filter and equalizer should be designed so as to make this term zero. In the following sections we will examine methods designed to achieve this.

Exercise 3.1

The regenerator equalizer output waveform in a digital transmission system is represented by

$$y(t) = \sum_{k=-\infty}^{\infty} A_k p_r(t - t_d - kT_b) \tag{3.7}$$

with $t_d = T_b/2$ and with $A_k = -1, 0$ or $+1$ volts. The basic received pulse shape $p_r(t)$ is as shown in Figure 3.3(a). Note that Equation (3.7) implies that noise and crosstalk are negligible since $n_o(t)$ is zero.

Find the value of the ISI components at each sampling instant if the input HDB3 code sequence $\{a_k\}$ represents a long string of alternating zeros and ones as shown in Figure 3.3(b).

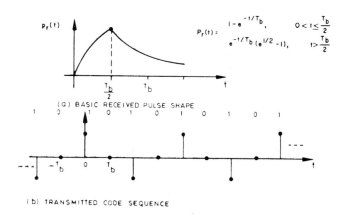

Figure 3.3 Example of ISI calculation.

Solution. The received waveform $y(t)$ will be the sum of pulses of the form shown in Figure 3.4.

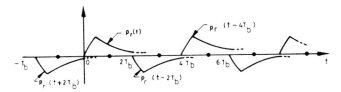

Figure 3.4 Constituent pulses in $y(t)$.

The optimum sampling instants are at

$$t = \ldots, 0.5T_b, 1.5T_b, 2.5T_b, \ldots.$$

(1) Firstly we consider sample times when the signal term is zero, that is, intervals in which a 0 is transmitted. For example, consider the ISI at sampling instant $t_m = 3.5T_b$. The ISI component is the sum of the ISI components from the tails of preceding pulses. That is, at $t = 3.5T_b$ the ISI is

$$y_I(3.5T_b) = -p_r(1.5T_b) + p_r(3.5T_b) - p_r(5.5T_b) + \ldots.$$

Now substituting the expressions for the basic pulse shape $p_r(t)$ from Figure 3.3(a) we obtain

$$\begin{aligned}
y_I(3.5T_b) &= (e^{\frac{1}{2}} - 1)\,(-e^{-1.5} + e^{-3.5} - e^{-5.5} + \ldots) \\
&= (e^{\frac{1}{2}} - 1)\,(-e^{-1.5})\,(1 - e^{-2} + e^{-4} \ldots) \\
&= (e^{\frac{1}{2}} - 1)\,(-e^{-1.5})\,(1/(1 - e^{-2})) \\
&= -0.167 \text{ volts.}
\end{aligned}$$

This will be the value of the ISI terms at each of the sampling instants $t = \ldots, -0.5T_b, 3.5T_b, 7.5T_b, \ldots.$

It is easy to see that at sampling times $t = \ldots, 1.5T_b, 5.5T_b, 9.5T_b, \ldots$ the ISI terms are each

$$y_I(5.5T_b) = +0.167 \text{ volts.}$$

(2) Next we consider sample times in symbol intervals when a 1 is transmitted. For example, the ISI at $t = 4.5T_b$ is from Equation (3.5)

$$\begin{aligned}
y_I(4.5T_b) &= -p_r(2.5T_b) + p_r(4.5T_b) - p_r(6.5T_b) + \ldots \\
&= (e^{\frac{1}{2}} - 1)\,(-e^{-2.5} + e^{-4.5} - e^{-6.5} + \ldots) \\
&= (e^{\frac{1}{2}} - 1)\,(-e^{-2.5})\,(1/(1 - e^{-2})) \\
&= -0.062 \text{ volts.}
\end{aligned}$$

This will be the ISI term at each of the sampling instants $t = \ldots , 0.5T_b,$ $4.5T_b, 8.5T_b, \ldots$. Note that the wanted signal term will be

$$1 - e^{-0.5} = +0.393 \text{ volts}$$

representing a $+1$ transmitted.

Likewise, the ISI term at $t = \ldots , 2.5T_b, 6.5T_b, .10.5T_b, \ldots$ will be -0.062 volts.

The wanted signal term will be

$$-0.393 \text{ volts}$$

representing a -1 transmitted.

In Exercise 3.1, the ISI values were calculated for the case where the transmitted a_k sequence is $\ldots -1\ 0\ 1\ 0\ -1\ 0 \ldots$. If the sequence was changed, then this would affect the amount of ISI that results. In order to examine worst-case effects, it would be necessary to consider all possible a_k patterns prior to the sampling instant under examination. The pattern length must be sufficient to include all pulses whose tails contribute significantly to the resultant ISI. If this number of pulses is large, then the number of ISI calculations required may become prohibitively large.

3.2 NYQUIST PULSE SHAPING

It is of considerable importance to know under what conditions we can transmit at a rate $r_b = 1/T_b$ symbols per second through a low-pass channel with *zero intersymbol interference* effects. It is clear from Equation (3.6) that for zero ISI, the received basic pulse $p_r(t)$ after equalization should be zero at all sampling instants other than the sampling instant when $p_r(t)$ is a maximum. That is, we require the transmit filter, channel transfer function, and receiver equalizer to be such that

$$p_r(kT_b) = \begin{cases} 1 \text{ for } k=0 \\ 0 \text{ for } k = \ldots -2T_b, -T_b, T_b, 2T_b, \ldots \end{cases} \tag{3.8}$$

In this section, we discuss techniques for achieving this.

3.2.1 Maximum Rate Pulses

One pulse shape that satisfies Equation (3.8) is the sinc pulse sinc $r_b t$ given by

$$p_r(t) = \frac{\sin \pi r_b t}{\pi r_b t} \tag{3.9}$$

which is shown plotted in Figure 3.5. The Fourier Transform of this pulse provides its spectral content. From Appendix 2.1, it can be seen that the transform is constant for frequencies from 0 to $1/2T_b$ (Hz), and zero elsewhere.

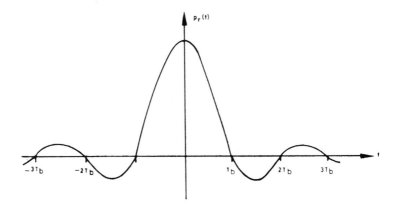

Figure 3.5   The sinc pulse.

The following theorem is well known.

Theorem 3.1: Nyquist-rate Signalling Theorem

If r_b independent symbols per second are to be transmitted in an ideal low-pass channel of bandwidth B (Hz), the maximum rate of $r_b = 2B$ symbols/sec can only be achieved without ISI if the regenerator basic pulse shape $p_r(t)$ is a sinc pulse as specified in Equation (3.9).   ■

The signalling rate $r_b = 2B$ is known as the Nyquist rate. Then when Nyquist rate signalling is used, the received line waveform will be

$$y(t) = \sum_{k=-\infty}^{\infty} A_k \operatorname{sinc}(r_b t - kT_b). \tag{3.10}$$

A typical waveform for this case was shown in Figure 2.7. Unfortunately, this basic pulse shape cannot be generated in practice. That is, it is noncausal. The existence of its infinite length "leading tails" implies that it would require a filter with infinite delay to generate the pulse. Even if we could generate sinc pulses, the regenerator performance would be critically dependent on timing accuracy since any small error in sampling instants could lead to performance deterioration. We must therefore seek more practical schemes for achieving high bit rates in a given bandwidth.

3.2.2 Symbol Packing Rate

In comparing different pulse shaping schemes for zero ISI, it is convenient to compare them in terms of a *symbol packing rate*. This is expressed in terms of the number of symbols per second per hertz (sym./s/Hz) that is achievable (at least in principle) without ISI. For example, if HDB3 symbols were transmitted at the maximum rate $r_b = 2B$, as discussed in Theorem 3.1, then the packing rate will be 2 bits/s/Hz. In practice, transmission rates below the Nyquist rate must be used. Next we will examine a method for doing this.

3.2.3 Nyquist Vestigial Symmetry Criterion for Zero ISI

It is possible to find a whole class of pulse shapes which satisfy Equation (3.8) and which can therefore be used to achieve transmission with zero ISI. The following theorem indicates how the condition for zero ISI can be expressed as a constraint on the Fourier Transform $P_r(f)$ of the basic pulse $p_r(t)$. This condition is called the Nyquist pulse shaping vestigial symmetry criterion.

Theorem 3.2: Nyquist Vestigial Symmetry Theorem

If $P_r(f)$ satisfies

$$\sum_{k=-\infty}^{\infty} P_r(f + k/T_b) = K \text{ for } |f| < \frac{1}{2T_b} \qquad (3.11)$$

with K an arbitrary constant, then $p_r(t)$ produces zero ISI since it satisfies Equation (3.8). ■

The proof of this theorem is not given here. See for example Shanmugan (1). At first sight, Equation (3.11) may appear a little difficult to interpret. Note that the criterion of Equation (3.11) does not uniquely specify the spectral shape $P_r(f)$. What then is its significance?

It may clarify the interpretation of Equation (3.11) if we choose for illustration a specific $P_r(f)$ function that satisfies the criterion. Consider for example the spectral shape shown in Figure 3.6(a). It should be noted that this is *not* a practical $P_r(f)$ function but is chosen here simply to explain the significance of Equation (3.11).

Expanding out Equation (3.11) we obtain

$$\ldots P_r(f - 1/T_b) + P_r(f) + P_r(f + 1/T_b) + \ldots = T_b \text{ for } |f| < 1/T_b.$$

For the $P_r(f)$ function shown in Figure 3.6(a), it is apparent that only the above three terms contribute to the sum in the region where $|f| < 1/2T_b$. Figures 3.6(b) and (c) show the other two shifted functions which must be considered.

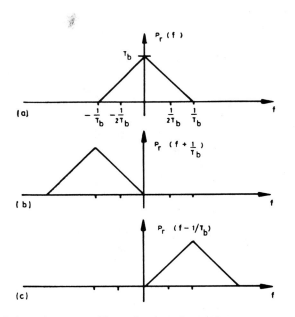

Figure 3.6   Pulse spectra illustrating Nyquist vestigial symmetry criterion.

Clearly their sum is equal to T_b in the interval $|f| < \dfrac{1}{2T_b}$ so Equation (3.11) is satisfied.

How can we describe the class of spectral shapes $P_r(f)$ which satisfies Equation (3.11)? In summary, the regenerator basic pulse $p_r(t)$ must have a Fourier Transform $P_r(f)$ for which the positive frequency part is 6 dB down at $f = \dfrac{1}{2T_b}$ and which has odd symmetry about that point (hence the term "vestigial symmetry"). This is illustrated in Figure 3.7.

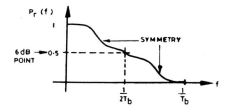

Figure 3.7   Illustration of vestigial symmetry for zero ISI.

There are of course an infinite number of spectra and associated pulse shapes which satisfy this requirement. In the next section, we examine a class of such spectral shapes which are of practical significance.

3.2.4 Raised Cosine Spectrum for Zero ISI

We have seen that the Nyquist criterion of Equation (3.11) does not uniquely specify the spectral shape $P_r(f)$ of the basic pulse required at the equalizer output. In practice, $P_r(f)$ must be chosen bearing in mind two important considerations.

(1) *filter complexity*—The shape of $P_r(f)$ determines the ease with which shaping filters can be realized. A smooth roll-off characteristic is preferable to one with discontinuities.

(2) *rate of decay of $p_r(t)$*—A basic pulse $p_r(t)$ with a fast rate of decay and smaller values in the vicinity of adjacent sampling instants $\pm T_b$, $\pm 2T_b$, . . . is desirable. This ensures that small timing errors do not cause ISI.

An acceptable compromise is the class of $P_r(f)$ functions with a raised cosine frequency characteristic (illustrated in Figure 3.8). Note that $P_r(f)$ consists of a flat amplitude region and a roll-off region that has sinusoidal form. The width of the flat region is variable. Figure 3.8 shows plots of $P_r(f)$ (for positive frequencies only). The various functions are specified in terms of a variable α which determines the bandwidth occupied by the pulse spectrum $P_r(f)$.

The raised cosine spectrum functions are given by

$$P_r(f) = \begin{cases} K & , |f| \leq \dfrac{1}{2T_b}(1-\alpha) \\[2mm] K\cos^2 \dfrac{\pi T_b}{2-\alpha}\{|f| - \dfrac{1}{2T_b}(1-\alpha)\} & , \dfrac{1}{2T_b}(1-\alpha) < |f| \leq \dfrac{1}{2T_b}(1+\alpha) \\[2mm] 0 & , |f| > \dfrac{1}{2T_b}(1+\alpha) \end{cases} \tag{3.12}$$

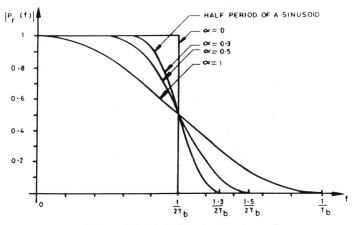

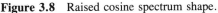

Figure 3.8 Raised cosine spectrum shape.

where $0 < \alpha \leq 1$.

From Figure 3.8, it is clear that a choice of α near to 0 would minimize the bandwidth requirement. That is, it would ensure that the symbol packing rate is high. It is important to examine the associated basic regenerator pulse shape $p_r(t)$, the Inverse Fourier Transform of $P_r(f)$. This is given by

$$p_r(t) = \frac{\cos \pi \alpha t}{T_b \left\{ 1 - \left(\dfrac{2 \alpha t}{T_b} \right)^2 \right\}} \; \frac{\sin \pi r_b t}{\pi r_b t} \tag{3.13}$$

which is shown in Figure 3.9.

Now we can summarize the factors that must be considered in the selection of the parameter α.

(1) *Symbol packing rate*—From Figure 3.8, it is apparent that the price we pay for choosing a smoother spectral shape (larger α) is reduced symbol packing rate (bandwidth efficiency). These rates vary between a maximum of two and a minimum of one symbol/s/Hz as shown in Table 3.1.

(2) *Bit timing effects on ISI*—By choosing larger values of α, the basic pulse shape $p_r(t)$ will have smaller residual "ripple" components beyond T_b seconds on either side of its peak value. That is, the pulse decays away more rapidly. Faster decaying pulses mean that synchronization will be less critical. Modest bit timing errors will not cause large amounts of ISI.

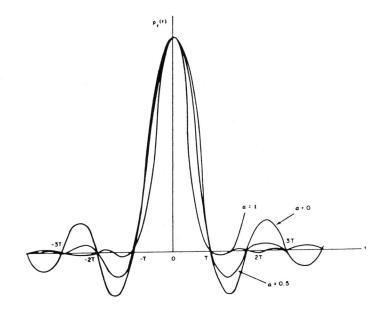

Figure 3.9 Nyquist Pulse for four values of parameter α.

Table 3.1
Symbol packing rates for various Nyquist pulses.

α	symbol packing rate symb/s/Hz	bandwidth occupied (Hz)
0	2.0	$0.5/T_b$
0.3	1.54	$0.65/T_b$
0.5	1.33	$0.76/T_b$
1.0	1.0	T_b

It should be noted that, strictly speaking, none of the raised cosine pulse spectra is physically realizable because of the preringing and postringing that theoretically extends out infinitely in time. In practice, Figure 3.8 represents a goal that can only be approximated by practical filtering.

We note that the received pulse shape $p_r(t)$ is a function of the transmitted waveform pulse shape $p_T(t)$, the channel characteristics, and the receiver equalizer filter. In terms of Figure 3.1, we have

$$P_r(f) = K_C P_T(f) H_C(f) H_R(f) e^{-j2\pi f t_d} \tag{3.14}$$

where K_C is a constant and t_d represents the time delay in the transmission channel.

In principle, the transmitted pulse may have a raised cosine spectrum and the channel $H_C(f)$ and equalizer $H_R(f)$ perfectly "flat" responses. Alternatively, the channel and equalizer may be designed to have a raised cosine transfer function, and the transmitted pulse a flat spectrum (narrow pulses). Often the practical situation is somewhere between these two cases. In practice, many 1.5 and 2 Mbit/s line systems use "half-width" rectangular pulses. The transmitted basic pulse $p_T(t)$ is a rectangular pulse of width $T_b/2$.

The transmission channel transfer function is likely to vary with time. Changes in transmission line characteristics result particularly from temperature changes. The equalizer therefore may need to be adaptively altered to compensate for these changes to keep the ISI to a minimum. We will examine the design of such equalizers in a later section. Next, we will review some techniques for designing shaping filters which approximate the Nyquist raised-cosine gain-frequency response characteristics.

3.2.5 Pulse Shaping Circuits

Consider the case where the raised-cosine spectrum shaping is to be implemented at the transmitter. The design of the pulse shaping filter may be carried out by one of two methods.

(1) *Analogue Filter Technique*—A network with a raised cosine gain-frequency response and a linear phase response may be approximated by conventional filter design techniques. For example, Feher (2) provides detail on a seventh-order phase-equalized elliptic filter which approximates the Nyquist raised

cosine roll-off with $\alpha = 0.3$. It is suggested that it is sufficient to satisfy the linear phase requirement up to the 10 to 15 dB attenuation point.

(2) *Digital Filter Technique*—An alternative method of generating pulse waveforms with prescribed time functions makes use of a binary transversal filter as shown in Figure 3.10.

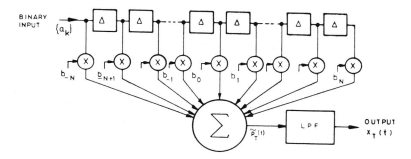

Figure 3.10   Pulse shaping using a transversal filter.

The encoded impulse sequence $\{a_k\}$ is fed into a tapped delay line consisting of $2N$ delay elements each of delay Δ(secs). For a ternary signalling scheme, each delay element must consist of two shift registers.

The constants $b_{-N}, b_{-N+1}, \ldots, b_0, b_1, \ldots, b_{N-1}, b_N$ are multiplier constants which determine the impulse response of the filter. Consider a single positive symbol $+V$ entering the filter at time $t = -N$ followed by a long string of 0's. Let the filter clock pulses shift the input $+V$ pulse through the delay line at a rate of several times faster than the data bit rate. In practice, the number of memory elements $2N$ may be typically 20-30.

Then the transversal filter output $\tilde{p}_T(t)$ will be in the form of the staircase waveform shown in Figure 3.11. In that figure, the solid curve represents the desired impulse response shape $p_T(t)$ of the filter. The filter constants

$$b_{-N}, b_{-N+1} \ldots, b_N$$

are sample values of $p_T(t)$. The output $\tilde{p}_T(t)$ is the sum

$$\tilde{p}_T(t) = \sum_{j=-N}^{N} b_j s(t - \Delta j) \qquad (3.15)$$

where

$$s(t) = \begin{cases} 1 & , \quad -\Delta/2 < t < \Delta/2 \\ 0 & , \quad \text{otherwise.} \end{cases}$$

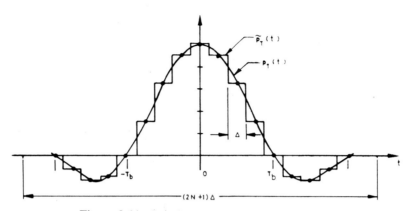

Figure 3.11 Impulse response of transversal filter.

The output $\tilde{p}_T(t)$ is smoothed with a simple low-pass filter. In the case of a single $+V$ "impulse" input followed by zeroes, the output approximates the required pulse shape $p_T(t)$.

In the example shown, the input symbol is shifted through the shift register at four times the data rate since we have used $4\Delta = T_b$. For a general input, successive a_k bits are shifted in once every T_b seconds and the filter forms the convolution of the input sequence and the impulse response of the filter. That is, the output composite waveform $x_T(t)$ represents the superposition of successive pulses that result from successive input bits in the sequence $\{a_k\}$.

Exercise 3.2

Consider a digital shaping filter with nine delay units and with multiplier tap values given as follows

b_{-4}	b_{-3}	b_{-2}	b_{-1}	b_0	b_1	b_2	b_3	b_4
0	-0.1	0	0.6	1.0	0.6	0	-0.1	0

Draw up a table of output values from the analogue summer at times Δ, 2Δ, 3Δ, . . . , 15Δ for the following input sequence

$$\{a_k\} = \{1\ 0\ 0\ 0\ -1\ 1\ 0\ 0\}$$

Solution. The table of values is as follows:

Time	Input	Delay line contents									Output
t	a	-4	-3	-2	-1	0	1	2	3	4	$x_T(t)$
Δ	1	1	—	—	—	—	—	—	—	—	0
2Δ	—	—	1	—	—	—	—	—	—	—	-0.1
3Δ	0	0	—	1	—	—	—	—	—	—	0
4Δ	—	—	0	—	1	—	—	—	—	—	0.6
5Δ	0	0	—	0	—	1	—	—	—	—	1.0*
6Δ	—	—	0	—	0	—	1	—	—	—	0.6
7Δ	0	0	—	0	—	0	—	1	—	—	0 *
8Δ	—	—	0	—	0	—	0	—	1	—	-0.1
9Δ	-1	-1	—	0	—	0	—	0	—	1	0 *
10Δ	—	—	-1	—	0	—	0	—	0	—	0.1
11Δ	1	1	—	-1	—	0	—	0	—	0	0 *
12Δ	—	—	1	—	-1	—	0	—	0	—	-0.7
13Δ	0	0	—	1	—	-1	—	0	—	0	-1.0*
14Δ	—	—	0	—	1	—	-1	—	0	—	0
15Δ	0	0	—	0	—	1	—	-1	—	0	1.0*

*Regenerator sample instants.

3.3 MULTILEVEL SIGNALLING

In Section 3.2 we saw that it is possible, at least in principle, to transmit data symbols in an ideal low-pass channel at a rate up to $2B$ symb/s without ISI. If a ternary code such as AMI or HDB3 is used, then each three-level symbol represents one binary bit. In that case, we can achieve bit packing rates up to 2 bit/s/Hz. Other codes, such as the 4B3T or the 5B4T codes, achieve higher bit packing rates.

Theorem 3.1 does not just apply to binary symbols. In fact, the amplitude of the symbols being transmitted over the Nyquist-rate signalling channel can have any real value. Since other pulses contribute zero ISI at receiver sampling instants, the relative amplitude of each symbol will be recovered exactly at the receiver if noise is not significant. This will be true even though, after transmission, the pulses associated with individual symbols overlap into other symbol intervals. At the sampling instants their ISI is zero.

Multilevel signalling schemes are those in which each transmitted symbol may take on one of M levels. Often M is chosen as a power of 2, for example, $M = 4$, 8, or 16. These schemes are also called M-ary signalling schemes. When $M = 4$, it is sometimes referred to as *quaternary signalling*.

Consider for example, an encoder and transmit filter for a quaternary scheme ($M = 4$). This is illustrated in Figure 3.12. The input binary data stream is sub-divided into blocks with two bits in each. Then each encoder output symbol represents $\log_2 M = 2$ binary input symbols.

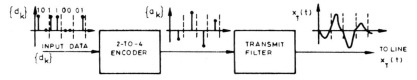

Figure 3.12 Quaternary ($M = 4$) transmission.

In multilevel signalling, each symbol represents $\log_2 M$ bits of information. Using raised-cosine spectrum pulses, it is possible to signal at rates r_S without ISI in a bandwidth B (Hz) with signalling rates in the range

$$B < r_S < 2B.$$

The symbol rate r_s is given the unit of ''bauds.'' If the M-ary symbol signalling rate is r_s bauds, this is equivalent to a bit rate of

$$r_b = r_S \log_2 M \qquad \text{(bit/s)}. \qquad (3.16)$$

This means that with four-level transmission, it is possible to transmit without ISI at a rate up to 4 bit/s/Hz. Similarly, with $M = 8$, rates up to 6 bit/s/Hz are possible and with $M = 16$, a symbol packing rate up to 8 bit/s/Hz is possible without ISI.

Multilevel signalling schemes with $M = 4$, 8, or 16 are seldom used for base-band digital transmission even though they can be potentially more bandwidth efficient. The main reason is related to the resultant deterioration in bit-error rates that inevitably follow (for the same transmitter power). This is because the regenerator has to sample the received signal and compare its amplitude against several more closely spaced amplitude thresholds. In the presence of noise, ISI, and poor synchronization, regenerator error probabilities increase rapidly as the number of signal levels increases.

For a specified bit-error rate, there is a tradeoff between bandwidth and signal-to-noise ratio (and hence, transmitter power). For example, consider systems with $M = 2$, 4, and 16 levels, respectively, with raised cosine ($\alpha = 1$) signalling in the presence of white Gaussian noise. The input data rate r_b is maintained constant. In order to obtain a bit-error rate of 10^{-4}, the $M = 4$ scheme requires approximately 4 dB more average power at the transmit filter output than the $M = 2$ scheme. The $M = 16$ scheme requires approximately 14 dB more transmitter power than the binary scheme. See, for example, Shanmugan (1) for detail.

Although M-ary signalling schemes are not often used in baseband line transmission, they are of considerable importance in digital radio systems and also for data transmission over analogue telephone channels. These will be considered in Chapter 1, Volume 2.

Multilevel signalling has also been used in the American AT&T system for the transmission of high-speed digital signals over line transmission plant. See Bylanski (3) for details.

3.4 CORRELATIVE (PARTIAL-RESPONSE) SIGNALLING

So far we have considered how pulse shaping techniques and transmission rates are related for a given channel bandwidth. It appears that in practical systems, it is necessary to signal at rates considerably below the theoretical limit of $2B$ symb/s if ISI is to be avoided.

In the mid-1960s Lender suggested that the independent signalling assumptions associated with the Nyquist criteria were unnecessarily restrictive (2). Nyquist signalling is based on the premise that each digit is transmitted independently. That is, an input symbol is confined to influencing the regenerator sample value in its own time slot. Adjacent symbol pulses must somehow be forced or shaped to have zero effect on the sample value in that time slot. The Nyquist assumption that the ISI at the sampling points can be eliminated implies that the successive symbols are independent and uncorrelated. Such systems, are sometimes referred to as zero-memory systems. As we have seen, the Nyquist maximum signalling rate of two symb/s/Hz cannot be achieved in zero-memory systems because of practical filter realizability problems.

Correlative (partial response) signalling techniques encode in such a way that there is some correlation between data symbols. That is, they deliberately introduce a controlled amount of symbol interdependence (ISI) over a span of one, two, or more adjacent symbols. At the receiving end, the controlled ISI is taken into account in regeneration and decoding.

The results are remarkable. With correlative techniques the following benefits are obtained:

(1) *High transmission rates*—Nyquist rate signalling is achievable in practice. Practical filter schemes can be realized to permit transmission at rates of $2B$ symb/s/Hz.

(2) *Timing perturbation tolerance*—Line waveforms in correlative schemes compare favorably with those of zero-memory schemes in their ISI tolerance to clock timing perturbation.

(3) *Speed tolerance*—Speed tolerance is commonly used as a measure of the sensitivity of a digital transmission system to changes in signalling rate, especially when rates exceed the Nyquist limit. Correlative encoding allows Nyquist rates to be exceeded by 15–20 percent. In zero-memory systems, error rates build up rapidly because of unavoidable ISI if the Nyquist rate is exceeded, even if no noise is present.

Historically, the first correlative encoding scheme was the duobinary signalling technique devised by Lender in 1962. Subsequently, other related coding techniques have been suggested. The term "partial response" was introduced by some researchers in 1965 to represent the general class of correlative encoding schemes. Two schemes that we will examine in detail are

(1) duobinary signalling
(2) class-4 partial response signalling, known also as modified duobinary signalling.

As a preliminary, an explanation of some of the terms may be helpful.

Correlative—implies finite memory with interdependence between symbols.

Partial response—a term synonomous with correlative.

Duobinary—a term which derives from "doubling the speed of binary."

3.4.1 Elementary Duobinary Scheme

Consider the system shown in Figure 3.13. It is substantially the same block diagram as used to represent zero-memory baseband transmission except that the first element consists of a simple first-order digital filter.

Let us examine the operation of the components in Figure 3.13.

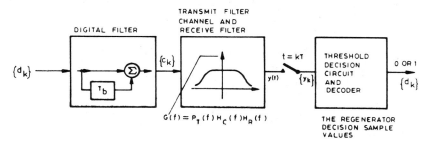

Figure 3.13 Components in a duobinary scheme.

Digital Filter

The input to the digital filter is a binary polar data sequence, such as, for example,

$$\{d_k\} = (-1\ -1\ +1\ -1\ +1\ +1\ -1\ -1\ +1).$$

The digital filter output symbol c_k is formed as the sum of the current input data value and the previous value. That is, the kth output symbol is

$$c_k = d_k + d_{k-1} \tag{3.17}$$

where $c_k \in \{-2, 0, 2\}$.

Note that the $\{c_k\}$ symbols are now no longer independent. Each symbol carries information which is partially "contained in" the preceeding and succeeding symbols.

In order to arrive at the transmission pulse shapes, we need to know the transfer function of the digital filter. The output is the sum of signals from two paths, one a direct connection and the other through a delay element. Therefore, the filter transfer function $H_d(f)$ can be represented by

$$H_d(f) = 1 + 1.e^{-j2\pi f T_b}. \tag{3.18}$$

This is illustrated in Figure 3.14.

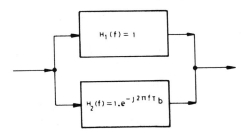
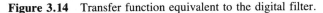

Figure 3.14   Transfer function equivalent to the digital filter.

In order to plot the gain-frequency response $|H_d(f)|$ we rewrite Equation (3.18) as

$$H_d(f) = e^{-j\pi f T_b} \{e^{j\pi f T_b} + e^{-j\pi f T_b}\}$$

Since

$$\cos A = \frac{1}{2}(e^{jA} + e^{-jA}) \tag{3.19}$$

we obtain

$$H_d(f) = e^{-j\pi f T_b} (2 \cos \pi f T_b). \tag{3.20}$$

The first term in the above is just a phase term whose magnitude is 1 so the gain frequency response of the filter becomes

$$|H_d(f)| = |2\cos\pi fT_b|.$$ (3.21)

This is shown in Figure 3.15.

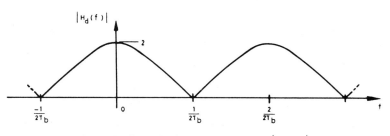

Figure 3.15 Gain frequency response $|H_d(f)|$.

Received Pulse Shapes

Next we consider the low-pass filtering represented by

$$G(f) = P_T(f)\,H_C(f)\,H_R(f)$$ (3.22)

associated with the transmit filter, channel and receiver filter (equalizer). For simplicity, let us assume that $G(f)$ represents an ideal low-pass filter with linear phase and with bandwidth $1/2T_b$. That is

$$G(f) = \begin{cases} T_b e^{j\pi fT_b}, & |f| \le \dfrac{1}{2T_b} \\ 0, & \text{otherwise.} \end{cases}$$ (3.23)

Then the transfer function of the system is

$$P_r(f) = H_d(f)\,G(f)$$

so that

$$P_r(f) = \begin{cases} 2T_b e^{-j\pi fT_b}\cos(\pi fT_b), & |f| \le \dfrac{1}{2T_b} \\ 0, & \text{otherwise.} \end{cases}$$ (3.24)

The inverse Fourier Transform of $P_r(f)$ gives the impulse response of the system. That is, it provides the shape of the received basic pulse $p_r(t)$ at the regenerator sampler input for a single positive data symbol d_k preceded and succeeded by long strings of zeros.

To find the inverse Fourier Transform of $P_r(f)$ we note that Equation (3.24) can be written

$$P_r(f) = [2T_b e^{-j\pi fT_b} \cos(\pi fT_b)] \, [G_1(f)] \tag{3.25}$$

where

$$G_1(f) = \begin{cases} 1, & |f| \le \dfrac{1}{2T_b} \\ 0, & \text{otherwise.} \end{cases}$$

Now the inverse Fourier Transforms of each of the bracketed terms are easy to find. For the first term, we use the Fourier Transform pair (denoted by $\leftrightarrow$) from Appendix 2.1.

$$2T_b \cos\pi fT_b \leftrightarrow T_b\delta(t - \frac{T_b}{2}) + T_b\delta(t + \frac{T_b}{2}).$$

Next we use the fact that multiplication by a complex exponential in the frequency domain is equivalent to a shift in the time domain. That is, if $X(f) \leftrightarrow x(t)$ then

$$X(f) \, e^{-j\pi fT_b} \leftrightarrow x(t - \frac{T_b}{2})$$

so that we obtain

$$2T_b \, e^{-j\pi fT_b} \cos(\pi fT_b) \leftrightarrow T_b\delta(t) + T_b\delta(t - T_b). \tag{3.26}$$

Also we have for the $G_1(f)$ function, the well known transform pair

$$G_1(f) \leftrightarrow \frac{1}{T_b} \operatorname{sinc} \frac{t}{T_b} . \tag{3.27}$$

Then we use the fact that convolution of two signals in the time domain corresponds to the multiplication of the Fourier Transforms in the frequency domain. From Equations (3.26) and (3.27) we obtain the inverse transform of Equation (3.25) as

$$P_r(f) \leftrightarrow [T_b\delta(t) + T_b\delta(t - T_b)] * \left[\frac{1}{T_b} \operatorname{sinc} \frac{t}{T_b} \right] \tag{3.28}$$

where * denotes the convolution operation. Finally, to perform this convolution we use the sifting property of the delta function, namely that for any function $g(t)$ we have

$$\delta(t - T_b) * g(t) = g(t - T_b). \tag{3.29}$$

As a result, we obtain the regenerator basic pulse response $p_r(t)$ corresponding to the above $P_r(f)$ as

$$p_r(t) = \frac{\sin \pi t/T_b}{\pi t/T_b} + \frac{\sin \pi(t - T_b)/T_b}{\pi(t - T_b)/T_b} . \tag{3.30}$$

This is the superposition of two sinc pulses, sinc (t/T_b) centered on $t = 0$ and sinc$(t - T_b)/T_b$ centered on $t = T_b$. Through the use of trigonometric manipulation we can also write $p_r(t)$ as

$$p_r(t) = \frac{4 \cos\{\pi(t - T_b/2)/T_b\}}{\pi\{1 - 4(t - T_b/2)^2/T_b^2\}} . \tag{3.31}$$

Plots of $P_r(f)$ and $p_r(t)$ are shown in Figure 3.16.

Note that $p_r(t)$ passes through zero at all sampling times except 0 and T_b. In comparison, zero memory Nyquist pulses are nonzero at sampling time 0 only.

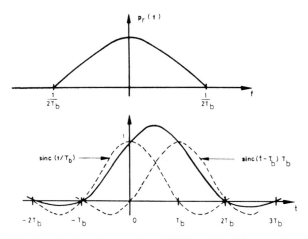

Figure 3.16 Receiver basic pulse spectrum $P_r(f)$ and duobinary pulse shape $p_r(t)$.

Exercise 3.3

Consider the above duobinary system with polar input data sequence

$$\{d_k\} = (-1 \ -1 \ +1 \ -1 \ +1 \ +1 \ -1 \ -1 \ +1)$$

Find the digital filter output sequence $\{c_k\}$ and sketch the approximate received waveform $y(t)$.

Solution. Using Equation (3.17), we obtain the values shown in Table 3.2 (where we have assumed that the previous value d_0 was -1). Note that the duobinary encoding scheme has converted an independent binary sequence $\{d_k\}$ into a correlated ternary sequence $\{c_k\}$.

To obtain $y(t)$, the waveform at the output of the regenerator equalizer, we superimpose sequences of basic pulses $p_r(t)$ of the form shown in Figure 3.16.

Table 3.2
Duobinary scheme sequences.

k	1	2	3	4	5	6	7	8	9
$\{d_k\}$	-1	-1	$+1$	-1	$+1$	$+1$	-1	-1	$+1$
$\{c_k\}$	-2	-2	0	0	0	$+2$	0	-2	0

That is

$$y(t) = \sum_k d_k \, p_r(t - kT_b) \tag{3.32}$$

giving the result shown in Figure 3.17.

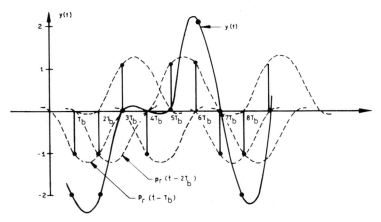

Figure 3.17   Received duobinary waveform.

Note from Figure 3.17 that the waveform $y(t)$ takes the values specified by the sequence $\{c_k\}$ at sampling instants T_b, $2T_b$, ... In the absence of noise or distortion in the system, the receive sample values y_k will be equal to the transmitted values. That is

$$y_k = c_k \quad \text{(noiseless case)}. \tag{3.33}$$

3.4.2 Nonbinary Inputs

The correlative encoding concept can be extended from binary to multilevel (M-ary) inputs. When the input sequence has M levels with $M = 2^q$ say (q being a positive integer), the correlative encoding scheme produces $2M - 1$ levels. The following exercise illustrates a quaternary-to-7-level encoding.

Exercise 3.4

The following unipolar binary data sequence $\{d_k\}$ is the input to a quaternary-to-7-level coder.

$$\{d_k\} = (1\ 1\ 0\ 1\ 0\ 0\ 0\ 0\ 1\ 0\ 1\ 1\ 1\ 1)$$

List the quaternary and correlative encoded symbols.

Solution. The quaternary sequence $\{b_k\}$ is found by simple binary-to-quaternary conversion. The correlative encoded sequence $\{c_k\}$ is found using

$$c_k = b_k + b_{k-1}.$$

The 7-level c_k sequence is shown in Table 3.3 (where we have assumed that the value of the b_k input previous to the first in the table is 0).

3.4.3 Duobinary Decoding and Error Propagation

Now we return to the duobinary scheme with binary inputs as illustrated in Table 3.2. We have seen that in the ideal noiseless case, the regenerator sample values y_k will be the same as the c_k values. Given the received samples $y_k = c_k$, how does the decoder of Figure 3.13 recover the output data symbols $\tilde{d}_k$? It can simply perform the inverse of Equation (3.17), namely estimate the kth data bit $\tilde{d}_k$ using

$$\tilde{d}_k = y_k - \tilde{d}_{k-1}. \tag{3.34}$$

For example, consider the decoder operation to find $\tilde{d}_4$ given the sequence shown in Table 3.2. If the previous decoder output value is assumed to be $\tilde{d}_3 = +1$, then

$$\tilde{d}_4 = c_4 - 1 = -1.$$

This decoding process is simple and would work well provided the channel is noiseless. However, if channel noise or ISI causes the regenerator to make

Table 3.3
Quarternary-to-7-level encoding.

$\{d_k\}$	1	1	0	1	0	0	0	0	1	0	1	1	1	1
$\{b_k\}$		3		1		0		0		2		3		3
$\{c_k\}$		3		4		1		0		2		5		6

an error so that $\tilde{d}_k \neq d_k$ for some k, then error propagation occurs. To illustrate, consider the decoding operation just described for finding $\tilde{d}_4$ in Table 3.1. Now assume that $\tilde{d}_3 = -1$. That is, an error occurred in regenerating and decoding $\tilde{d}_3$ (say a negative noise spike reduced y_3 from 0 to approximately -2). Assume $y_k = c_k$ for all $k \geq 4$, that is, no further errors occur. Then using the decoding rule specified by Equation (3.34) the decoder would obtain,

$$\tilde{d}_4 = 0 - (-1) = +1 \qquad (\neq d_4)$$

$$\tilde{d}_5 = 0 - (+1) = -1 \qquad (\neq d_5)$$

$$\tilde{d}_6 = +2 - (-1) = +3 \qquad (\neq d_6)$$

and so on. That is, the error in $\tilde{d}_3$ results in the propagation of additional errors through the subsequent decisions.

3.4.4   Precoding

The remedy for error propagation is to use a precoder as shown in Figure 3.18. Also shown in Figure 3.18 is a unipolar-to-bipolar convertor. This is followed by the duobinary system as described above.

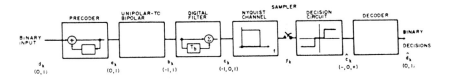

Figure 3.18   Duobinary system with precoding.

The precoder consists of a modulo-2 adder (exclusive-OR) circuit and a one-bit delay element in a feedback path. For a binary input d_k, the kth precoder output is

$$a_k = d_k + a_{k-1} \qquad (3.35)$$

where a_k and d_k is each either 0 or 1. That is, a_k is the sum of the current data input and the previous precoder output, the addition being modulo-2.

Next, the unipolar a_k sequence is converted to a bipolar sequence b_k with values -1 or $+1$. (Recall that, as illustrated in Table 3.1, we assumed a bipolar input to the duobinary encoder.) The unipolar-to-bipolar conversion can be represented by the rule

$$b_k = 2a_k - 1 \qquad (3.36)$$

where normal algebraic operations are used.

Then as a result of the duobinary digital filter and Nyquist channel we obtain from Equation (3.17) and (3.33) that for an error-free channel, the kth regenerator sample value is

$$y_k = b_k + b_{k-1}. \tag{3.37}$$

Exercise 3.5

Let the data input sequence $\{d_k\}$ given in Exercise 3.4 be the input to the complete duobinary system shown in Figure 3.18. Tabulate the sequence values $\{a_k\}$, $\{b_k\}$, and $\{y_k\}$ assuming an ideal noiseless channel.

Solution. The values shown in Table 3.4 are obtained using Equations (3.35), (3.36), and (3.37) assuming that $a_0 = 0$.

Table 3.4
Duobinary system sequences using a precoder.

k	0	1	2	3	4	5	6	7	8	9	10	11	12	13
$\{d_k\}$	—	1	1	0	1	0	0	0	0	1	0	1	1	1
$\{a_k\}$	0	1	0	0	1	1	1	1	1	0	0	1	0	1
$\{b_k\}$	-1	1	-1	-1	1	1	1	1	1	-1	-1	1	-1	1
$\{y_k\}$	—	0	0	-2	0	2	2	2	2	0	-2	0	0	0

3.4.5 Regeneration and Decoding for the Duobinary Scheme with Precoding

Now we can determine the operation required by the regenerator decoding circuit for the complete duobinary scheme including the precoder.

(1) *Ideal channel case*—If no noise or distortion occurs in transmission, the regenerator sample values $\{y_k\}$ will be as illustrated in Table 3.4. Compare the d_k and y_k values in the table. It is obvious that the decoder operation is such that it chooses the data estimate $\tilde{d}_k$ using

$$\tilde{d}_k = \begin{cases} 0 & \text{if } y_k = -2 \text{ or } +2 \\ 1 & \text{if } y_k = 0. \end{cases} \tag{3.38}$$

(2) *Practical channel case*—In practice some noise and ISI may occur in transmission. In this situation, the regenerator samples the received waveform $y(t)$ to obtain sample y_k. The sample value y_k may now have any real value. The decision circuit and decoder ignore the sign of the sample y_k. They

simply compare the magnitude of y_k against a threshold V_T set midway between the expected 0 and +2 values.

Then the decoder rule is

$$\tilde{d}_k = \begin{cases} 0 & \text{if} \quad |y_k| \geq V_T \\ 1 & \text{if} \quad |y_k| < V_T. \end{cases} \tag{3.39}$$

We can now highlight some important characteristics of the duobinary scheme.

(1) *Error propagation*—We observe from Equations (3.38) and (3.39) that each data bit is decoded independent of the history of previous bits, despite the correlation properties. No knowledge of any sample other than y_k is used in determining $\tilde{d}_k$. As a result, no error propagation can occur. The precoding scheme has overcome the error propagation effect in the elementary duobinary system.

(2) *Starting conditions*—In Exercise 3.5, it was assumed that the initial value a_0 of the precoder output is $a_0 = 0$. It would be a considerable disadvantage if it were necessary for the initial state of the precoder to be known at the regenerator. We study this in Exercise 3.6.

Exercise 3.6

Repeat Exercise 3.5 for the same input sequence but assume that the initial precoder output value is $a_0 = 1$.

Solution. The details are left to the reader. The resultant output sequence $\{y_k\}$ is as tabulated below

$\{d_k\}$	1	1	0	1	0	0	0	0	1	0	1	1	1
$\{y_k\}$	0	0	2	0	−2	−2	−2	−2	0	2	0	0	0

Now examine these sequences in comparison with those in Table 3.4. It can be seen that the decoder rules given in Equations (3.38) and (3.39) still apply. That is, the decoder operation is independent of the initial value a_0 and this information is not required at the receive end. The precoder has solved this problem as well as the error-propagation problem.

(3) *Precoding for M-ary inputs*—For M-level inputs, the precoder required is specified by

$$a_k = d_k + a_{k-1} \qquad (\text{modulo-}M). \tag{3.40}$$

(4) *Low frequency spectrum of duobinary signal*—So far we have concentrated on the high frequency characteristics of the line signal power

spectrum which rolls off to a maximum value of $B = 1/2T_b$. We should also pay attention to the low frequency content of the signal $y(t)$ as we did for all the zero-memory codes previously studied in Chapter 2. Consider, for example, an input data sequence consisting of a long string of zeroes. Then the duobinary sequence $\{y_k\}$ (noiseless case) would consist of a long string of $+2$ volts (for $a_0 = 0$). This indicates that for a duobinary signal

(i)   dc and low frequency line spectrum components are not attenuated. With ac coupled transmission systems, baseline wander would be a problem.

(ii) for long sequences of marks or spaces, no bit interval transitions are available for synchronization recovery.

A number of other duobinary schemes have been suggested which result in line spectra without dc and significant low frequency components. To summarize these, we first examine a generalization of the duobinary scheme, namely generalized correlative encoding. Then we examine one important subclass of these schemes, the modified duobinary scheme, also known as Class-4 partial response encoding.

3.4.6   Generalized Correlative Encoding

Other correlative encoding schemes can be developed which have line signals with different spectral density characteristics. Figure 3.19 shows a generalized correlative encoding arrangement.

A transversal filter is used for encoding. This consists of a tapped delay line and tap value multipliers with integer valued constants $h_0, h_1, \ldots, h_N$. By choosing various values for the constants $h_0, h_1, \ldots, h_N$, different encoding configurations are obtained with different degrees of correlation between sym-

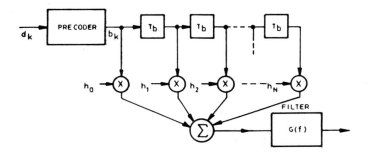

Figure 3.19   Generalized correlative encoding.

bols. Also, different signal spectra are obtained as a result. The duobinary encoding discussed above represents the case where $N = 1$ and $h_0 = h_1 = 1$.

For each generalized correlative encoding scheme, an appropriate precoder is required. Its structure will be related to the encoding scheme as illustrated in Section 3.4.4. The precoder ensures that, at the receiver, independent bit decisions can be made based on single samples of the received waveform. This avoids the problem of error propagation.

3.4.7 Modified Duobinary (Class-4 Partial Response) Scheme

Consider the configuration of Figure 3.19 with $N = 2$ delay units and with

$$h_0 = 1, \ h_1 = 0 \text{ and } h_2 = -1.$$

The resultant configuration is the modified duobinary encoder scheme (also known as a Class-4 partial response encoder). It has a correlation span of two bit intervals.

It follows that for this scheme, the digital filter transfer function is

$$H_d(f) = 1 - e^{-j4\pi f T_b}$$

$$= 2j e^{-j2\pi f T_b} \sin 2\pi f T_b$$

so that the received signal spectrum is

$$P_r(f) = \begin{cases} 2j e^{-j2\pi f T_b} \sin 2\pi f T_b & \text{for } |f| < \dfrac{1}{2T_b} \\ 0 & \text{otherwise.} \end{cases} \tag{3.41}$$

For this case, the impulse respone $p_r(t)$ obtained by the inverse Fourier Transform of $P_r(f)$ becomes

$$p_r(t) = \frac{\sin \pi t/T_b}{\pi t/T_b} - \frac{\sin \pi(t - 2T_b)/T_b}{\pi(t - 2T_b)/T_b}. \tag{3.42}$$

Both $P_r(f)$ and $p_r(t)$ are shown plotted in Figure 3.20. As can be seen, the spectrum has nulls at dc and at the Nyquist frequency ($1/2T_b$). The basic pulse has an average value of zero.

In order to implement the modified duobinary filtering scheme described by Equation (3.41), it is useful to note that $H_d(f) = 1 - e^{-j4\pi f T_b}$ can also be written

$$H_d(f) = (1 - e^{-j2\pi f T_b})(1 + e^{-j2\pi f T_b}). \tag{3.43}$$

The term inside the first brackets can be implemented using the duobinary-type digital filter. The second term represents the cosine roll-off associated with

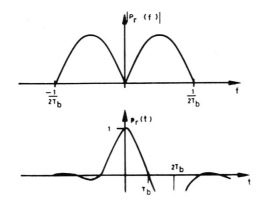

Figure 3.20   Spectrum and impulse response of the modified duobinary scheme.

the conventional duobinary scheme. As such it can be implemented using an analogue cosine filter such as one of a switched-capacitor type. (Switched-capacitor filters are discussed in Chapter 6, Volume 1).

Performance compared with duobinary scheme: Both the duobinary and the modified duobinary schemes generate three-level sequences. They are commonly employed in digital radio systems (to be discussed in Chapter 1, Volume 2). They have also been used to advantage for baseband line transmission for PCM. See, for example, Pasupathy (5).

The modified duobinary scheme does not require transmission of dc and low frequencies. However, in practice, it does not usually perform quite as well as the duobinary scheme when high transmission rates are involved. The ISI in a practical modified duobinary scheme is likely to be worse than that for the duobinary scheme (assuming some filtering imperfections). This is because in the duobinary scheme, the only transitions permitted in successive bits are between two adjacent amplitude levels. Such a restriction does not exist for the modified duobinary waveforms, so increased ISI can result from imperfect filtering or timing errors. See for example, Lender's Chapter 7 in Feher (2) for details.

Correlative encoding schemes represent a useful approach in systems where bandwidth efficiency (bit packing rate) has to be maximized. In general, however, the bit error rate performance of three-level duobinary schemes will not be quite as good as for zero-memory three-level schemes operating at bit rates below the Nyquist rate. Bit error rates will be examined in the next chapter.

Alternative approaches have been proposed to obtain minimum bandwidth line coding schemes. For example, see Kim (6).

3.5 PROBLEMS

3.1 (1) A rectangular pulse of height 1 (V) and width T_b (s) is transmitted through a first-order low-pass channel, with transfer function

$$H_c(f) = 1/(1 + jf/B).$$

Obtain an expression for the received pulse shape $p_r(t)$.

(2) Show that when $B = 1/T_b$, $p_r(t)$ will be as given in Figure 3.3.

3.2 Consider the received pulse described in Problem 3.1. Find the peak value of $p_r(t)$ for the cases where

(1) $B = 1/T_b$

(2) $B = 2/T_b$.

If $p_r(t)$ is the basic received pulse shape in an HDB3 encoded waveform, at what voltage level should the regenerator decision threshold values be set if intersymbol interference and noise are ignored?

3.3 At the output of a regenerator equalizer in a baseband transmission system the basic received pulse shape $p_r(t)$ is as shown in Figure 3.3. Assume that the binary input system is encoded before transmission using an AMI code. The regenerator samples each data symbol in the center of the T_b interval. Consider the intersymbol interference (ISI) component of the received signal at a given sampling instant t'. Assume that up to N preceding symbols can give rise to significant ISI at $t = t'$. Determine the input binary sequence of those N symbols which results in the most severe ISI for the cases where $N = 3, 4, 5$, respectively.

3.4 Consider the ''basic pulse'' shape at the output of a regeneration equalizer to be given by

$$p_r(t) = \frac{\sin \pi t/T_b}{\pi t/T_b}.$$

Assume that the received pulse sequence A_k is periodic and of the form

$$\ldots + 1, 0, -1, 0, +1, 0, -1, \ldots.$$

The received waveform $y(t)$ is sampled with a timing error of one-tenth of a bit duration. That is, $y(t)$ is sampled at

$$t_m = mT_b + T_b/10.$$

Find the value of the intersymbol interference (ISI) associated with samples t_0 and t_1, respectively for two cases:

(1) include only the ISI that results from symbols in the three adjacent symbol intervals before and after t_o

(2) include the ISI from all symbol intervals.

3.5 (1) Assuming the use of raised-cosine Nyquist shaping, obtain an expression for the maximum pulse transmission rate in terms of the transmission bandwidth B and the roll-off factor α.

(2) Find also an expression for the symbol packing rate in terms of α.

(3) If M-level signalling is used where $M = 2^n$, n an integer, find an expression for the bit packing rate in terms of α and n.

3.6 An analogue source is sampled, quantized, and encoded into a binary code using eight bits per sample. The binary pulses are encoded using an HDB3 code and transmitted over a bandwidth of 20 kHz using Nyquist shaping with $\alpha = 0.5$.

(1) Find the maximum possible binary pulse rate.

(2) Find the maximum permissible source bandwidth.

3.7 In an ideal duobinary transmission system including precoding, the input binary sequence is

$$\{d_k\} = (0\ 1\ 0\ 0\ 1\ 1\ 1\ 0\ 0\ 1\ 0\ 1).$$

Find the two possible corresponding sets of regenerator sample sequences $\{y_k\}$.

3.8 Repeat Problem 3.7 for the modified duobinary scheme.

3.9 Consider a duobinary transmission system including precoding which is operating over a noisy channel. The regenerator decisions are based on sample values at the midpoint of signalling symbols. The regenerator produces the following sequence of ternary decisions

$$+\ +\ 0\ 0\ 0\ -\ 0\ 0\ -\ 0\ +\ 0\ 0\ -\ 0\ +\ +\ 0\ -$$

(1) Can you determine whether a transmission error has occurred?

(2) Assuming only one error has occurred, find all possible estimates of the transmitted sequence.

3.10 It is claimed that certain error patterns can be detected in a duobinary system and that, in the absence of errors, the regenerator decisions c_k must obey the following rule:

> The values of two successive c_k symbols at the extreme levels $+2$ or -2 must be different if the number of intervening bits at the level 0 is odd. Otherwise, they must be the same.

Show whether or not this is correct.

3.11 Referring to Question 3.10, determine an equivalent error detection rule for the modified duobinary scheme.

Hint: Divide the y_k sequence into two trains of odd-numbered and even-numbered sequences.

3.12 For a quaternary to 7-level correlative encoding scheme, what is the probability distribution of the seven levels? Assume that for the input binary data $P(1) = P(0) = 0.5$.

3.6 REFERENCES

1. K. S. Shanmugan, *Digital and Analogue Communication Systems*, Wiley, 1979.
2. K. Feher, *Digital Communications—Microwave Applications*, Prentice-Hall, 1981.
3. P. B. Bylanski and D. G. W. Ingram, *Digital Transmission Systems*, Peter Peregrinus, 1980.
4. G. J. Semple, "The Effect of Intersymbol Interference on the Operation of PCM Line Regenerators," *ATR Australian Telecommunication Research*, Vol. 12, No. 1, pp. 17–31, 1978.
5. S. Pasupathy, "Correlative Coding: A Bandwidth-efficient Signalling Scheme," *IEEE Communications Society Magazine*, Vol. 15, No. 4, pp. 4–11, July 1977.
6. D. Y. Kim and Jo-K. Kim, "A condition for stable minimum-bandwidth line codes," *IEEE Trans. on Comm.*, Vol. COM-33, No. 2, pp. 152–157, Feb. 1985.

Chapter 4

SIGNAL REGENERATION

4.1 INTRODUCTION

In the previous chapter, we examined various alternative signalling schemes for baseband transmission of digital signals. Our interest centered on signal types for which the intersymbol interference (ISI) could either be made zero (zero-memory systems) or else some well controlled quantity (correlative encoding systems). In each case, we considered the signal characteristics particularly in terms of the received waveform after passing through the receive amplifier and equalizer filter of the regenerator.

In this chapter, we examine the regenerator operation in more detail. One of the main advantages of digital transmission over traditional analogue systems lies in the regenerator performance capabilities. That is, providing the line system is designed so that the noise and ISI is properly controlled, then the regenerator output bit error rate can be made very low. Cascading many regenerative repeater sections together, therefore, does not lead to system degradation with circuit length. In analogue repeater systems, noise effects accumulate with circuit length. Also, when analogue transmission circuits consist of several cable pairs inside a common sheath, crosstalk from other circuits can be a severely limiting factor. Digital transmission is potentially capable of providing a greater number of derived telephone circuits in a cable than frequency-division-multiplex (FDM) analogue techniques. This is because the regenerators are capable of good performance in environments with higher crosstalk interference.

In the following sections, we will examine the factors that affect regenerator bit error rate (BER) performance in baseband digital transmission. The equalizer in the regenerator plays a vital role in compensating for channel distortion and ensuring that signalling waveforms are provided to the regenerator decision circuits with minimum undesirable ISI. Several approaches to equalizer design will be discussed.

Clock recovery procedures are important in self-timing regenerators to ensure that the received equalized waveform is sampled at optimum sampling instants.

In this chapter, we will consider systems for recovering bit-timing information at the regenerator. The timing accuracy may be affected by the signal bit-pattern characteristics as well as by noise and other effects. It is important to understand how these various sources of timing errors can have a significant effect on overall BER performance.

The regenerator uses decision circuits to estimate from samples of the received waveform which digital symbol was transmitted. In this chapter we will explore the factors that affect the BER performance of these decision circuits in the presence of noise (which includes crosstalk) and ISI.

4.2 REGENERATIVE REPEATERS

4.2.1 Functions

A regenerator performs the process of recognizing and reconstructing a received digital signal so that the resultant amplitude, waveform shape, and timing are as near as possible to the original transmitted values. The regenerator functions include equalization, clock recovery, waveform sampling, level comparison, pulse reconstruction, transmit filtering, and line amplification. In a digital transmission system consisting of several repeater sections, the regenerative repeaters perform a number of ancillary functions in addition to the regenerator process.

Figure 4.1 shows a block diagram of a typical regenerative repeater. Figure 4.2 shows some typical waveforms. The waveforms $v_A(t)$, $v_B(t)$, and $v_C(t)$ are typical of those obtained at points A, B, and C in the regenerator of Figure 4.1.

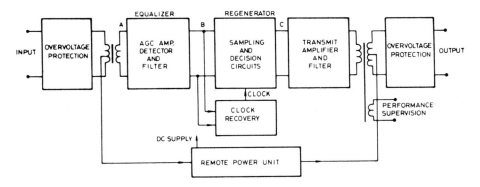

Figure 4.1 Block schematic of a regenerative repeater.

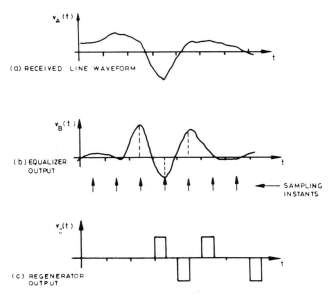

Figure 4.2 Typical regenerator waveforms.

In addition to its primary function as a regenerator, the regenerative repeater also performs the following functions:

(1) *Overvoltage protection*—from lightning and other external sources.
(2) *Remote power feeding*—A dc voltage in the range 40–200 volts is supplied on the cable pair from the terminal and is used to drop the required 10–15 volts across each repeater.
(3) *System supervision*—Information about the bit stream recovered at the regenerator is usually made available for return to the transmit terminal via a voice-frequency analogue supervisory circuit connected to each repeater in the circuit. One method commonly used to check each regenerator in a circuit is to send a series of test sequences down the line. Each test signal in the sequence consists of a ternary sequence containing a particular low frequency periodic component. Each regenerator is fitted with a monitoring bandpass filter with center frequency unique to that regenerator. The regenerated test signal is filtered and the filter output returned to the terminal via the supervisory circuit. The terminal may progressively send a series of test signals with different low frequency components and check out continuity in each successive regenerative repeater. This will be discussed in more detail in Chapter 5, Volume 1.

Often the regenerative repeaters are constructed in housings suitable for underground location. Circuit design that uses microcircuit miniaturization and requires the minimum of manual adjustments is essential. For the regenerator in Figure 4.1, the circuit for the equalizer might consist of one VLSI circuit chip, and the regenerating circuit including clock recovery (but not tuned circuits) another VLSI circuit.

4.2.2 Clock Recovery

It is necessary for each regenerator to incorporate a circuit which performs symbol (or bit) synchronization. This circuit processes the waveform from the equalizer output to identify the relative starting point for each symbol interval. The recovered symbol synchronization signal is necessary for the timing of all other regenerator functions including sampling, level comparison, and transmit pulse shaping.

A number of different approaches to symbol synchronization are possible. Consider the following possibilities:

(1) A differentiator and a controlled oscillator might be synchronized to sudden transitions in the received signal. The differentiator would provide a pulse at each sharp signal amplitude transition between $-1, 0$, and 1 levels. The oscillator (clock) could then be designed to oscillate at approximately the symbol rate and be synchronized to the differentiation pulses.

(2) A narrow band filter or narrow band phase-lock loop could be used to extract the symbol clock frequency component from the line signal. A phase-lock loop could be used to regenerate a symbol clock in synchronism with that in the received waveform.

The above linear processing techniques may not perform well for some types of baseband transmission signals. The first technique relies on the received waveform containing sharp transitions between the signalling levels. Examination of waveform $v_B(t)$ in Figure 4.2 clearly illustrates that the equalized waveform may not contain these transitions. The second technique can be used when the received signal spectrum contains a strong component at the symbol frequency $f = r_b$. Reference to Figure 2.26 shows that for many line coded waveforms, the spectra do not contain any frequency component at $f = r_b$. For example, AMI and HDB3 coded line waveforms have a null at the clock frequency. In this case, linear timing extraction techniques cannot be used.

In practice, therefore, the symbol timing extraction circuit in a regenerator usually contains a nonlinear signal processing subunit. Two possibilities exist:

(1) A zero-crossing detector (comparator) could be used to initiate the generation of a timing pulse used to control the frequency of a voltage controlled oscillator (VCO). Such a scheme is illustrated in Figure 4.3. Whenever the

received waveform passes through zero with positive slope say, the comparator and one-shot generates a pulse of duration approximately equal to $r_b/2$, that is, one half a bit interval. The phase detector compares the timing of this half-width pulse with a square wave output from the VCO. The phase detector output is a dc voltage proportional to the phase difference. This is used to control the VCO frequency. If the VCO signal drifts out of phase, the VCO control voltage derived from the phase comparator output attempts to correct the VCO frequency. Details of such a system are given in Shanmugan (6).

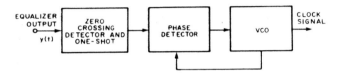

Figure 4.3   Clock recovery using a zero-crossing detector.

(2) A full-wave rectifier and threshold detector can be used. This is a very popular method for clock recovery for HDB3 and AMI line coded signals. The clock recovery scheme is illustrated in Figure 4.4. When the equalizer output waveform is passed through the full-wave rectifier and threshold detector, the rectifier nonlinearity causes the spectrum of the output signal to contain a discrete frequency component at twice the symbol rate, that is at $f = 2r_b$ (despite the fact that there was a null at $f = r_b$ in the original signal $y(t)$). The following elements of Figure 4.3(b) are then used to filter out the component at $f = 2r_b$ and use it to regenerate a clock signal at the received symbol rate r_b. Further details of this technique can be found in Bylanski (7).

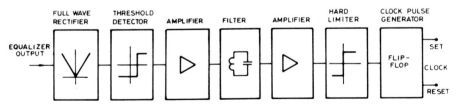

Figure 4.4   Bit-timing recovery.

The principles of synchronization techniques are discussed extensively in Lindsey (8), Spilker (9), and Stiffler (10).

In practice, as a result of noise and other effects, it is not possible for clock recovery circuits to avoid some random fluctuation of the bit-timing. This is referred to as *timing jitter*. One of its effects is to cause the timing of the regenerated pulses to fluctuate about their correct positions. Unfortunately, jitter may accumulate along a chain of regenerators and can have a major effect on overall system performance.

4.2.3 Sampling and Decision Circuits

In the sampling and level decision circuits, the equalized pulse waveform is sampled at instants which ideally are $1/T_b$ (secs) apart and occur at times corresponding to the occurrence of symbol peaks. Next, the decision circuits must estimate which transmit symbol is represented by each sample value.

Most of the line codes discussed in Chapter 2 consist of ternary symbols. That is, received sample values may represent one of three decision values, say $+V$, 0 and $-V$ (volts). Therefore, the regenerator decision circuits must compare each sample value against two threshold voltage values to determine which of the three symbol values is the best estimate of the transmitted symbol that was sent in that symbol interval.

Figure 4.5(a) illustrates in block schematic form a typical regenerator decision system and pulse shaping circuit for a ternary coded system. Figure 4.5(b) illustrates the waveforms and voltage decision values involved.

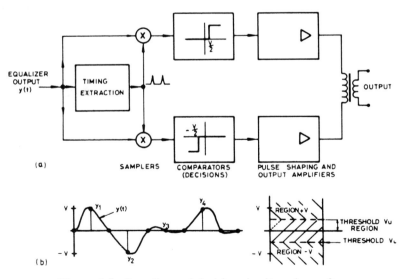

Figure 4.5 Sampling and decision circuits and waveforms.

The equalized waveform $y(t)$ is split into two parallel paths before being sampled once each symbol interval. In Figure 4.5 the sampling process is carried out by multiplying $y(t)$ with a train of narrow unit amplitude clock pulses. The resultant samples pass into comparators (slicers) with threshold voltages set at V_U and V_L, respectively. Typically, these threshold values will be set to approximately $V/2$ and $-V/2$ where V is the mean value of the pulse peaks in $y(t)$. If a given sample value exceeds V_U, then the comparator output voltages cause the output amplifiers to generate a current pulse in a given direction through the primary of the coupling transformer. On the other hand, if the received sample value is less than V_L, a current pulse is generated in the opposite direction through the coupling transformer. If the sample lies between the two thresholds (indicating a 0 symbol) then no current pulse flows into the transformer.

In many PCM repeater systems, the relative positions of the samplers and the comparators are reversed. That is, slicing is done before sampling. It is not difficult to see that this change of positions does not significantly affect the principle of the operation of the system. In that case, the inputs to the pulse shaping filters and output amplifiers will be in the form of narrow pulses.

The practical design of the decision circuits must take into account possible comparator imperfections such as:

(1) *threshold uncertainty effects*—they may result from a finite-slope comparator characteristic and may reduce the operating margin of the system.

(2) *hysteresis effects*—the effect of hysteresis on regenerator performance is not obvious. At first thought, it might appear that hysteresis effects would be advantageous in reducing the regenerator sensitivity to unwanted noise and result in improved BER performance. It turns out that the opposite is true. Hysteresis will result in degradation of BER performance. See Bylanski (7) for details.

4.3 EQUALIZATION

4.3.1 Functions

A vital element in determining the performance of a regenerator is the input amplifier and equalizer. This is required to amplify and filter the distorted incoming pulse train into a form that enables the decision circuits to choose the correct symbol sequence with minimum probability of error.

The automatic-gain-control (AGC) amplifier compensates for random variations in line attenuation. It attempts to maintain the peak positive or negative values of the equalizer output waveform at some predetermined levels. These levels must be set in conjunction with the values of the comparator thresholds in the decision circuits.

The equalizer circuit ideally should ensure that pulses with zero ISI are available at the input to the sampling and decision circuits. In addition, the equalizer reduces the effect of transmission line noise and crosstalk by filtering. Although this section of the receiver performs a filtering function, the term *equalizer*, is used rather than *filter*. This is because its characteristics must compensate for the pulse dispersion in time caused by the line section amplitude and phase distortion. The required equalizer output waveform should exhibit zero inter-symbol interference. Therefore, the time-domain response of the equalizer is of more critical significance than its frequency response.

In the following sections, we will examine in detail the principles of equalizer design to meet zero ISI requirements. To accommodate a wide range of line parameters and also possible variations with time, it is usually necessary to provide some means for varying the equalizer characteristics to match the line. We will also examine methods for varying the equalizer parameters adaptively.

Consider the transmission system block schematic shown in Figure 4.6. $H_R(f)$ represents the transfer function of the receive amplifier and equalizer. Note that in some systems, $H_R(f)$ may consist of a fixed part and a variable part. The fixed equalizer filter section is sometimes called a line built-out network. It is intended to provide zero ISI under average conditions. This is followed by a "mop-up" variable equalizer section which can be varied to compensate for small ISI variations with time as a result of circuit or transmission line variations. We will use $H_R(f)$ to represent the total receiver filtering whether or not it is all achieved in one equalizer circuit or in separate fixed and variable sections.

We assume the line code input in Figure 4.6 to be a sequence of impulses or delta functions of weight 1, 0, or -1 (representing either $+1$, 0 or -1 for a ternary line code). We consider here zero-memory coding systems, but the results can readily be extended to a correlative encoding system.

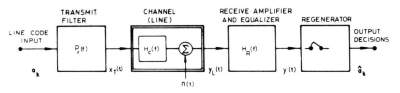

Figure 4.6   Transmission system schematic.

For a single positive pulse input, we require the equalizer $H_R(f)$ to ensure that the equalizer output pulse is zero at times spaced T_b(sec) intervals either side of the sampling point (pulse maximum). As in Chapter 3, for a single impulse input to the system (assumed noiseless), we write the equalizer output

$$y(t) = p_r(t - t_d)$$

where $p_r(t - t_d)$ is the impulse response of the transmission system, including the equalizer. The parameter t_d represents transmission delay.

Now we require the equalizer to ensure that the pulse response $p_r(t - t_d)$ causes zero ISI. This is illustrated in Figure 4.7. It can be seen that, for zero ISI, the pulse shape must satisfy

$$p_r(t - t_d) = \begin{cases} 1 & \text{for} \quad t - t_d = 0 \\ 0 & \text{for} \quad t = t_d + kT_b, \ k = \pm 1, \pm 2, \ldots \end{cases}$$

This is the equivalent to the requirement that

$$p_r(t) = \begin{cases} 1 & \text{for} \quad t = 0 \\ 0 & \text{for} \quad t = kT_b, \ k = \pm 1, \ \pm 2, \ldots \end{cases} \tag{4.1}$$

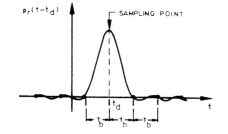

Figure 4.7 System impulse response for zero ISI.

In Chapter 3 we saw that there are a large number of pulse shapes which will satisfy equation (4.1).

Let $P_r(f)$ be the Fourier Transform of $p_r(t)$. Then by examination of Figure 4.6 it can be seen that

$$P_r(f) = P_T(f) H_C(f) H_R(f). \tag{4.2}$$

The required transfer function $H_R(f)$ of the equalizer can be determined once the transfer functions $P_T(f)$ and $H_C(f)$ of the transmit filter and channel are known. The transfer function $P_r(f)$ of the combined system given by Equation (4.2) must be chosen so as to satisfy Equation (4.1).

4.3.2 Transmit Pulse Shapes

In many practical digital line transmission systems, the transmitter output pulse shapes are rectangular. This is common practice, primarily because rectangular pulses are the easiest to generate. In many 2.048 and 1.544 Mbit/s PCM transmission systems, a half-width rectangular pulse is transmitted. That is, the impulse response $p_T(t)$ of the transmit filter is

$$p_T(t) = \begin{cases} V & \text{for } 0 < t \le T_b/2 \\ 0 & \text{otherwise.} \end{cases} \qquad (4.3)$$

Typically, $V = 3$ volts for 120 ohm balanced circuits such as junction cable pairs. On the other hand, $V = 2.37$ volts for 75 ohm unbalanced circuits such as for coaxial cables (11).

Instead of rectangular transmit pulses, smoother pulse shapes with reduced high frequency spectral content could, in principle, be expected to reduce crosstalk problems between pairs in the same cable sheath. Near end crosstalk (NEXT) is often the dominant crosstalk problem between cable pairs carrying PCM signals. NEXT refers to electromagnetic coupling from other digital transmitter lines into a regenerator cable pair where the transmitters and regenerator, respectively, are in the same location.

Later in this chapter we will examine another criteria for choosing the transmit pulse shape. If the noise and crosstalk at the regenerator can be modelled as white Gaussian noise, then it is possible to find the form that the pulse shaping function $P_T(f)$ should have to minimize the probability of error.

In practice, the additional complexity involved in obtaining complex transmit pulse shapes may not be justifiable. Rectangular pulse shapes can be much more readily obtained from the solid-state current amplifier needed to obtain the necessary cable input voltages.

4.3.3   Typical Transmission Line Characteristics

Before we examine methods for designing equalizer circuits to satisfy Equations (4.1) and (4.2), it may be helpful to examine some typical transmission characteristics encountered in practice.

Primary PCM (2.048 Mbit/s and 1.544 Mbit/s) systems are often used over urban junction cable pairs between exchanges. Symmetric pairs of 0.64 mm or 0.90 mm diameter are commonly used (1). Let $L(f)$ represent the gain (or loss) and phase versus frequency characteristics of the line. For the junction cable pairs described, it has been found empirically that the insertion loss $L(f)$ in dB of a plastic insulated cable section between 120 ohm terminations over the range 300 to 1500 kHz satisfies

$$L(f) = L(f_o) \{f/f_o)^{\frac{1}{2}}\} \qquad (4.4)$$

where $f_o = r_b/2$ is one half the bit rate. The parameter $L(f_o)$ is the insertion loss (in dB) at f_o, and f is the frequency in Hz. The line loss is plotted in Figure 4.8 for $L(f_o) = 20$ dB at $f_o = 1024$ kHz. Note, that in considering the overall channel loss, account must also be taken of the frequency response of coupling transformers, the location of which are shown in Figure 4.1. Regenerative repeaters for such junction cable systems are typically required to operate satisfactorily for insertion loss values $L(f_o)$ in the range 5–37 dB at 20°C.

Coaxial cables are also commonly used for baseband transmission, especially for high-speed transmission of higher-order multiplexed signals at speeds up to 140 Mbit/s. The attenuation in coaxial cables typically increases as the square root of frequency and can be found approximately (2) using

$$10 \log_{10} L(f) = 0.001 \ \alpha l \sqrt{(f/10^6)} \quad \text{(dB)} \qquad (4.5)$$

where $\alpha = 2.40$ for standard (2.6/9.5 mm) coaxial cables and $\alpha = 5.26$ for small (1.2/4.4 mm) coaxial cables. In the above equation, l is the cable length in meters and f is the frequency in Hz. A typical plot of attenuation for the standard coaxial cable is shown in Figure 4.8.

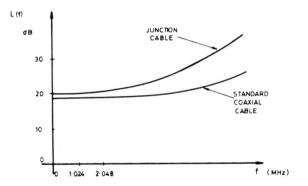

Figure 4.8 Typical line loss characteristics.

4.3.4 Frequency-Domain Characteristics of Equalizers

Given the characteristics of the transmit pulse and of the transmission channel, the equalizer can be designed to satisfy Equations (4.1) and (4.2). Two types of equalizer design are possible.

(1) *Analogue filter type of equalizer:* A filter network is designed to provide the transfer function $H_R(f)$ required to satisfy Equation (4.2). This can be referred to as a frequency-domain approach.

(2) *Transversal equalizer:* A tapped delay line transversal filter structure is designed to ensure the basic pulse shape $p_r(t)$ at the equalizer output satisfies Equation (4.1). That is, the basic pulse is a maximum at relative time $t = 0$ and is zero at instants $t = \pm T_b, \pm 2T_b, \ldots$. This can be referred to as a time-domain approach.

The following exercise illustrates the frequency-domain approach to equalizer design.

Exercise 4.1

Determine the equalizer transfer function $H_R(f)$ required for HDB3 transmission at 2048 kbit/s if half-width rectangular transmit pulses are used, and the channel transfer function is ideal low-pass with bandwidth 2048 MHz. the required received pulse shape $p_r(t)$ is to have a Nyquist raised-cosine spectrum (Equation (3.12)) with $\alpha = 1$.

Solution. From Equation (4.2), we require that the equalizer satisfy

$$H_R(f) = \frac{P_r(f)}{P_T(f)H_C(f)} \tag{4.6}$$

where from Equation (3.12)

$$P_r(f) = \begin{cases} K_1 \cos^2 \pi f T_b & 0 < |f| \leq 1/T_b \\ 0 & \text{otherwise.} \end{cases} \tag{4.7}$$

The basic transmit pulse is

$$p_T(t) = \begin{cases} V & , \quad 0 < t < T_b/2 \\ 0 & , \quad \text{otherwise} \end{cases}$$

with Fourier Transform from Appendix 2.1 given by

$$P_T(f) = \frac{VT_b}{2} \operatorname{sinc}(fT_b/2). e^{-j\pi f T_b/2}. \tag{4.8}$$

The transmit pulse spectrum, channel gain-frequency response, and required equalizer output spectrum are shown in Figures 4.9(a), (b) and (c).

From Equation (4.6) we obtain the desired equalizer transfer function as

$$H_R(f) = \begin{cases} \dfrac{k_1 \cos^2 \pi f T_b}{\dfrac{K_2 VT_b}{2} e^{-j\pi f T_b/2} \dfrac{\sin (\pi f T_b/2)}{\pi f T_b/2}} & \text{for } -1/T_b < f < 1/T_b \\ \\ 0 & \text{otherwise.} \end{cases} \tag{4.9}$$

The required gain-frequency response $|H_R(f)|$ of the equalizer is shown in Figure 4.9(d).

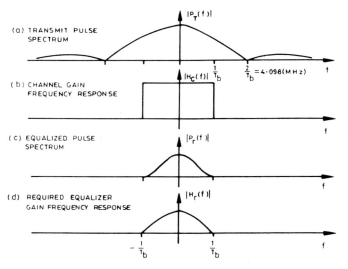

Figure 4.9 Transmission characteristics.

4.3.5 Transversal Equalizers

In the above example, the required equalizer was specified in terms of its frequency domain characteristic $H_R(f)$. An alternative approach is based on defining the required response in the time domain. After all, we are only interested in the output waveform values at a few prescribed sampling times. Since the received signal shape is only of interest at instants T_b seconds apart, a natural equalizer structure is a tapped delay line with taps spaced by T_b seconds, and with the tap outputs used to form weighted sums of signal samples T_b seconds apart. A tapped delay line equalizer structure is shown in Figure 4.10. This is also known as a transversal filter structure—hence, the term *transversal equalizer*.

The received line signal is designated $y_L(t)$ and the equalizer output $y(t)$. Consider the transmission system schematic in Figure 4.6. For an impulse (an isolated 1) at the transmitter input in Figure 4.6, we write the basic pulse shape at the channel input as $p_T(t)$, at the channel output as $p_L(t)$ and at the equalizer output as $p_r(t)$. The latter two are shown in parenthesis in Figure 4.10.

In practice, the equalizer delay line in Figure 4.10 may be constructed using groups of shift-registers. Then the input $y_L(t)$ must be sampled every T_b(secs) and converted to a digital signal using an A/D converter. Each sample value may, for example, be represented by a 10-bit binary number. Then, each T_b delay element shown would represent 10 shift registers clocked at rate $1/T_b$.

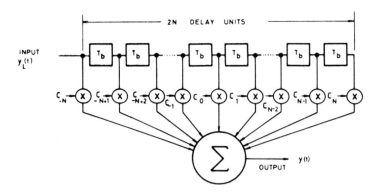

Figure 4.10   A transversal equalizer.

Alternatively, a charge-coupled delay (CCD) line structure may be used. In that case, the input $y_L(t)$ must be sampled once every T_b seconds. The voltage samples are then passed sequentially through the delay line at one shift every T_b seconds.

As the sample values are shifted at a rate of one shift every T_b(secs) through the delay line of $2N$ elements, each sample remains in the delay line for $2N.T_b$ seconds. The equalizer pulse shaping function will be determined by the $2N+1$ tap multiplier coefficients $C_{-N}, C_{-N+1}, \ldots, C_N$. These are chosen in such a way that the resultant $p_r(t)$ is zero at all sampling instants other than when the pulse is at its peak value.

For a pulse $p_L(t)$ at the equalizer input, we can write the output $p_r(t)$ as the weighted sum

$$p_r(t) = C_{-N}p_L(t) + C_{N+1}p_L(t - T_b) + \ldots +$$
$$C_0 p_L(t - NT_b) + \ldots + C_N p_L(t - 2NT_b). \quad (4.10)$$

It follows that we can write the equalizer input-output relationship as

$$p_r(t) = \sum_{n=-N}^{N} C_n p_L(t - (n+N)T_b). \quad (4.11)$$

If the equalizer input pulse $p_L(t)$ is assumed to have its center at $t=0$, the equalizer output can be assumed to reach its center at time $t = NT_b$. That is, the peak value arrives at the equalizer output NT_b seconds delayed when the pulse peak is in the center of the equalizer. We denote sampling times $t = t_k$ at the equalizer as

$$t_k = (k + N)T_b, \quad k = -N, -N+1, \ldots, N-1, N$$

so the center of the pulse is at the equalizer output when $k = 0$. This is illustrated in Figure 4.11.

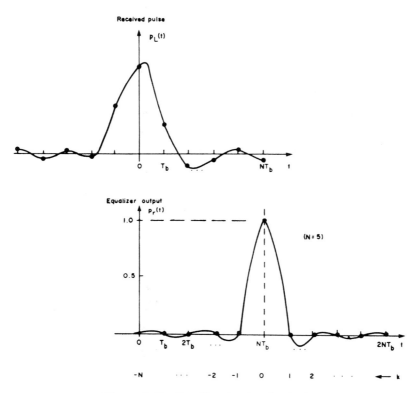

Figure 4.11 Equalizer input and output.

The sampled values of the input pulse at the $2N + 1$ equalizer tap points are from Equation (4.11):

$$p_L(t_k - (n + N)T_b) = p_L[(k + N)T_b - (n + N)T_b]$$

$$= p_L[(k - n)T_b]$$

for $n = -N, \ldots, N$. At sample time t_k we denote $p_L[(k-n)T_b]$ by $p_L(k-n)$.

Likewise, the equalizer output $p_r(t_k)$ is written by $p_r(k)$. Then equation (4.11) can be rewritten in terms of sample values as

$$p_r(k) = \sum_{n=-N}^{N} C_n p_L(k-n). \qquad (4.12)$$

Note that this is equivalent to discrete convolution of the received sample sequence $p_L(k)$ with the equalizer tap weights C_n. This is written in shorthand form

$$p_r(k) = C_n * p_L(k).$$

That is, the tap gains C_n determine the impulse response of the equalizer. Now we must determine the C_n values required for zero ISI.

Ideally, at the equalizer output we would like to have

$$p_r(k) = \begin{cases} 1 & \text{for} \quad k=0 \\ 0 & \text{for} \quad k=\pm 1, \pm 2, \ldots, \pm N. \end{cases} \tag{4.13}$$

An equalizer that is designed to achieve this is known as a *zero-forcing equalizer*. In a zero-forcing equalizer, we can only force $p_r(t)$ to zero at $2N$ sampling points since we have only $2N+1$ tap gain variables at our disposal.

Normally, the C_n values are greatest near the center of the equalizer and taper away to smaller values near the edges as the pulse dispersion effects reduce. In practice, N must be chosen in accordance with the amount of residual ISI tolerable at sample times greater than NT_b(secs) either side of the received pulse maximum ($k=0$). Typically, equalizers with up to 30–40 taps, are used.

The method of obtaining the C_n values is illustrated in the following exercise.

Exercise 4.2

For a three-tap equalizer ($N=1$), solve for the three C_n multiplier values given that the input basic pulse $p_L(t)$ has sample values

$$\{p_L(k)\} = (0, 0, 0.05, -0.15, 0.95, -0.2, 0, 0, 0).$$

Figure 4.12 shows the received line pulse at the equalizer input.

Solution. We require the C_n values to be chosen so that the three-tap equalizer output will have zero values at the two instants T_b either side of the output maximum.

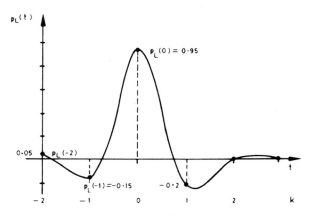

Figure 4.12   Sampled line output pulse.

From Equations (4.12) and (4.13) we have the simultaneous equations

$$0 = C_{-1}p_L(0) + C_0p_L(-1) + C_1p_L(-2)$$

$$1 = C_{-1}p_L(1) + C_0p_L(0) + C_1p_L(-1)$$

$$0 = C_{-1}p_L(2) + C_0p_L(1) + C_1p_L(0)$$

which we can solve for the C_n values given the $p_L(k)$ values. The solutions are

$$C_{-1} = 0.165$$

$$C_0 = 1.125$$

$$C_1 = 0.237$$

As a check, let us compute the equalized pulse values $p_r(k)$ using Equation (4.12). The values of $p_r(k)$ are shown below together with pre-equalized pulse values $p_L(k)$ for comparison.

$p_L(-3) = 0$	$p_r(-3) = 0.008$
$p_L(-2) = 0.05$	$p_r(-2) = 0.032$
$p_L(-1) = -0.15$	$p_r(-1) = 0$
$p_L(0) = 0.95$	$p_r(0) = 1.000$
$p_L(1) = -0.20$	$p_r(1) = 0$
$p_L(2) = 0$	$p_r(2) = 0.047$
$p_L(3) = 0$	$p_r(3) = 0$

Note that the equalized pulse has the required zero values for $k = \pm 1$. However, the equalized pulse has nonzero ISI values at $k = -3$ and 2 where the equalizer input had zero values. At these instants, the equalizer has had a detrimental effect. Of course, by increasing the number of delay elements $2N$, the number of zero ISI points in the equalized pulse can be increased.

Note that, unlike analogue equalizers for FDM systems where the transmission distortion is compensated at all frequencies, in transversal equalizers for digital transmission, compensation is only required at a finite number of time values.

Matrix form:
The set of simultaneous equations to be solved for the C_n values can be written more concisely by using the matrix form

$$(C_{-N}C_{-N+1} \ldots C_0 \ldots C_N) \begin{bmatrix} P_L(0) & P_L(-1) & p_L(-2N) \\ p_L(1) & p_L(0) & \\ & & \\ p_L(2N) & p_L(2N-1) & p_L(0) \end{bmatrix}$$

$$= (00 \ldots 010 \ldots 0) \qquad (4.14)$$

with N zeroes before the 1 and N zeroes after it.

In shorthand form, this can also be written

$$C.P_L = I \qquad (4.15)$$

where C is a row vector of tap gains, P_L is the square matrix for which the (i,j)th element is $p_L(i-j)$ and I is a $(2N+1)$ element row vector with the central element 1 and all others zero.

4.3.6 Automatic Equalizers

Automatic systems for setting up the equalizer tap gain values have been developed and are an important adjunct to most transversal equalizers. There are two classes of automatic equalizers.

(1) *Preset equalizers* are those in which the tap settings are adjusted by means of a known training sequence of test pulses prior to actual data transmission. The test sequence is transmitted and the tap gain values determined by observing the ISI that results. In one version of a preset equalizer, the training sequence consists of widely spaced 1 symbols separated by long strings of zero voltage values. In the equalizer, a feedback circuit adjusts the C_n tap gain values so that for each transmitted pulse the equalizer output has a maximum at the pulse center and minimum ISI values at sampling instants on either side. That is, the equalizer feedback circuit automatically adjusts the tap gains so that a zero-forcing response results. Then the test sequence is removed, the tap settings left unchanged and normal data is transmitted. Preset equalizers will only function correctly for a specific transmission line characteristic. Shanmugan (6) provides more details.

(2) *Adaptive equalizers* are designed such that the tap settings are continuously updated during data transmission in order to compensate for varying transmission line characteristics. We will examine these in more detail.

A general schematic for an adaptive equalizer is shown in Figure 4.13. Tap gains are adjusted up or down each bit interval based on an error signal

$$\epsilon_k = \bar{a}_k - y_k. \qquad (4.16)$$

As illustrated in Figure 4.13, $\tilde{a}_k$ is the output decision of the regenerator based on the equalizer output sample value y_k. When the equalizer is optimally adjusted, ϵ_k should be zero.

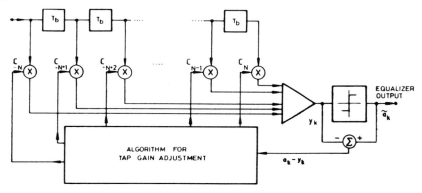

Figure 4.13   General form of adaptive equalizer.

In effect, this class of equalizers assumes that the regenerator decisions $\{\tilde{a}_k\}$ are correct and can be used in place of the transmitted sequence $\{a_k\}$ to adjust the equalizer adaptively. This type of equalizer is sometimes referred to as *decision-directed*.

Equalizer adjustment algorithms:

Considerable research has been undertaken on the criteria for adjusting the equalizer tap gain coefficients and on algorithms for optimizing the equalizer performance. Proakis (4) gives an excellent summary. Tap gain C_n values can be adjusted to minimize one of the following performance indices.

(1) *Bit-error rate* (P_e)—The average BER is the most meaningful performance measure as it takes both noise and ISI effects into account. However, it turns out that P_e is a highly nonlinear function of the C_n values. The function is difficult to compute, let alone minimize it. Consequently, this is not a practical performance index for C_n adjustment.

(2) *Peak distortion (zero-forcing)*—Using this performance measure, the equalizer attempts to minimize the sum of the magnitudes of the ISI terms $p_r(k)$ for $k \neq 0$ given by Equation (4.12). This is equivalent to the zero-forcing equalizer discussed above. The peak distortion algorithm has the disadvantage that it does not take the additive noise into account. Also the algorithm may not converge to a minimum if the initial ISI values are very high.

(3) *Mean square error criterion*—A performance index that does not suffer the limitations inherent in the other two criteria is the mean square error. This is given by

$$F(C_n) = E\{|\bar{a}_k - y_k|^2\} \qquad (4.17)$$

That is, the algorithm attempts to minimize the average (expected value) of the square of the magnitude of the difference between the kth data symbol and the kth equalizer output estimate of that symbol. This is known as the least mean square (LMS) algorithm originated by Widrow (4). It can be shown that one method of achieving this minimization is to adjust the tap gain coefficients adaptively using

$$C_n^{k+1} = C_n^k + \Delta(\bar{a}_k - y_k)r_n^{k+1}. \qquad (4.18)$$

That is, each tap value at time index $k+1$ is made equal to the previous tap value C_n^k modified by an amount equal to $\Delta(\bar{a}_k - y_k)r_n^{k+1}$. The parameter Δ is a positive constant and r_n^{k+1} is the sample value at the nth position of the equalizer delay line at time $k+1$. This is illustrated by a simple three-tap adaptive equalizer shown in Figure 4.14.

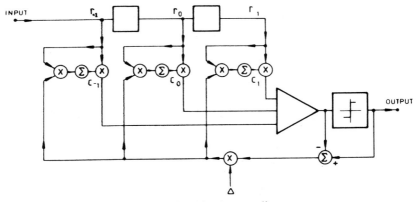

Figure 4.14   Adaptive equalizer.

In this type of equalizer the increment step size is proportional to the value of Δ. This must be chosen small enough to ensure convergence of the iterative procedure. However, if Δ is chosen too small, the equalizer may take longer to converge to optimum tap settings after initial switch-on or following any abrupt channel change. If Δ is not small enough, then once converged, the equalizer may generate excessive distortion because of responding to small noise values.

4.3.7 Computer Simulation

It is very helpful in the understanding of adaptive equalizer operation to simulate the equalization and tap setting algorithm in software. In this section, a simulation exercise is given in which the bit error rate (BER) performance is measured for a transmission system including an adaptive equalizer and regenerator decision circuit. This is particularly instructive in studying:

(1) the equalizer convergence characteristics during the initial stages of the data sequence (start-up period)
(2) the improvement in BER performance brought about by the equalizer
(3) the effect of varying degrees of ISI on BER performance
(4) the effect of channel noise on equalizer and regenerator performance
(5) the effect of varying the size of increments by which the tap settings are adaptively varied.

The block schematic in Figure 4.15 illustrates the system to be simulated. The simulation uses computer programs in Fortran language given in Table 4.1. Alternatively, a Basic language program given in Table 4.2 can be used. The reader is encouraged to simulate the system on a computer. The component elements in the simulation are as follows:

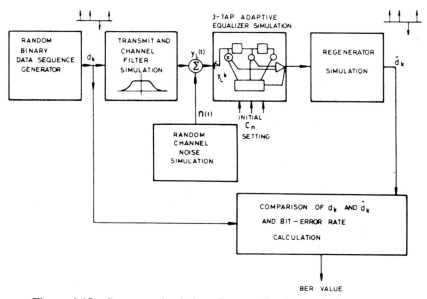

Figure 4.15 Computer simulation of an equalized transmission system.

Table 4.1
FORTRAN Simulation Program

```
00100         PROGRAM ADAPT
00110C
00120C        ADAPTIVE EQUALIZER SIMULATION PROGRAM
00130C        -------------------------------------
00140C        SUMMARY:                                      :
00150C        -------                                       :
00160C             THIS PROGRAM SIMULATES AN                :
00170C        ADAPTIVE EQUALIZER IN WHICH THE               :
00180C        TAP GAINS OF THE MULTIPLIERS ARE              :
00190C        CONTINUOUSLY UPDATED WITH THE USE             :
00200C        OF THE LEAST MEAN SQUARE ALGORITHM.           :

00230C        DEFINITION OF TERMS:                          :
00240C        --------------------                          :
00250C        DELTA: POSITIVE INCREMENTAL STEP              :
00260C               VALUE.                                 :
00270C        SNR: SIGNAL TO NOISE RATIO                    :
00280C        ATB: INTERSYMBOL INTERFERENCE                 :
00290C        SIGSQA : VARIANCE                             :
00300C        TWOPI: 6.283185307                            :
00310C        CMINUS: MULTIPLIER AT LEFT OF                 :
00320C                PULSE MAXIMUM                         :
00330C        CZERO: MULTIPLIER AT CENTRE OF THE            :
00340C               PULSE.                                 :
00350C        CPLUS: MULTIPLIER AT RIGHT OF THE             :
00360C               PULSE MAXIMUM.                         :
00370C        PE(I): PROBABILITY OF ERROR                   :
00380C        RAND1,RAND2 AND RAND3 ARE RANDOM              :
00390C        NUMBERS WHICH ARE GENERATED                   :
00400C        NOISE: RANDOM GAUSSIAN NOISE                  :
00410C        VOLTSQ: GAUSSIAN VARIANCE                     :
00420C        Y(K+1): OUTPUT                                :
00430C        R(K+1): OUTPUT + GAUSSIAN NOISE               :
00440C        SMPA(K+1): THE SAMPLED OUTPUT                 :
00450C        I,J,K: INTEGER VARIABLES                      :
00460C        FLAG: PULSE COUNTER                           :
00470C        A(K+1): INPUT PULSES                          :
00480C        DEXA(K+1): EXPECTED OUTPUT PULSES             :
00490C        ERRORS: ERROR COUNTER                         :
00500C        -------------------------------------
00510C
00520         REAL DELTA,SNR,ATB(4),SIGSQA,TWOPI,PE(4)
00530         REAL CMINUS,CZERO,CPLUS
00540         REAL RAND1,RAND2,RAND3,NOISE,VOLTSQ,Y(3),R(3),SMPA(3)
00550         INTEGER I,J,K,FLAG,A(3),DEXA(3),ERRORS,PLOT,N
00560         CHARACTER*59SPACE
00570         SPACE=' '
00580         TWOPI=8.0*ATAN(1.0)
00590         K=1
00600C
00610     5   PRINT 10
00620    10   FORMAT('INPUT FOUR ATB VALUES')
```

Table 4.1
FORTRAN Simulation Program—Continued

```
00630          DO 15 J=1,4
00640          READ*,ATB(J)
00650      15 CONTINUE
00660          PRINT 25
00670      25 FORMAT('INPUT DELTA VALUE')
00680          READ *,DELTA
00690C
00700C         INITIALIZATION OF ALL VARIABLES
00710C         - - - - - - - - - - - - - - - - - - - - - - - - - - - - -
00720          DO 45 J=1,4
00730          I=1
00740          DO 40 SNR=10,16,2
00750          SIGSQA=10**(-SNR/10)
00760          CMINUS=0
00770          CZERO=1
00780          CPLUS=0
00790          FLAG=0
00800          ERRORS=0
00810          Y(K)=0
00820          R(K)=0
00830          R(K+1)=0
00840C         - - - - - - - - - - - - - - - - - - - - - - - - - - - - -
00850C
00860C         RANDOM BINARY DATA SEQUENCE GENERATOR
00870C         - - - - - - - - - - - - - - - - - - - - - - - - - - - - - - - - - -
00880      20 RAND1=RANF()
00890          RAND2=RANF()
00900          RAND3=RANF()
00910          IF(RAND1.LE.0.5)THEN
00920              A(K+1)=-1
00930          ELSE
00940              A(K+1)=+1
00950          ENDIF
00960C
00970C         RANDOM CHANNEL NOISE SIMULATION
00980C         - - - - - - - - - - - - - - - - - - - - - - - - - - - - -
00990          VOLTSQ=-2.*SIGSQA*LOG(RAND2)
01000          NOISE=(SQRT(VOLTSQ))*COS(TWOPI*RAND3)
01010C
01020C         TRANSMIT AND CHANNEL FILTER
01030C         - - - - - - - - - - - - - - - - - - - - - -
01040          Y(K+1)=(A(K+1)*(1-(EXP(-ATB(J)/2.))))
01045+         +(Y(K)*(EXP(-ATB(J))))
01050          R(K+2)=Y(K+1)+NOISE
01060C
01070C         REGENERATOR SIMULATION
01080C         - - - - - - - - - - - - - - - - - -
01090          SMPA(K+1)=CMINUS*R(K+2)+CZERO*R(K+1)+CPLUS*R(K)
01100          IF(SMPA(K+1).GT.0)THEN
01110              DEXA(K+1)=1
01120          ELSE
01130              DEXA(K+1)=-1
```

Table 4.1
FORTRAN Simulation Program—Continued

```
01140        ENDIF
01150        IF(FLAG.LT.100) THEN
01160        GOTO 50
01170        ENDIF
01180        IF(DEXA(K+1).NE.A(K))THEN
01190             ERRORS=ERRORS+1
01200        ENDIF
01210C
01220C       3-TAP ADAPTIVE EQUALIZER SIMULATION
01230C       ------------------------------------
01240    50  CMINUS=CMINUS+DELTA*(DEXA(K+1)-SMPA(K+1))*R(K+2)
01250        CZERO=CZERO+DELTA*(DEXA(K+1)-SMPA(K+1))*R(K+1)
01260        CPLUS=CPLUS+DELTA*(DEXA(K+1)-SMPA(K+1))*R(K)
01270C
01280C       REFRESHING EQUALIZER VARIABLES
01290C       -----------------------------
01300        A(K)=A(K+1)
01310        Y(K)=Y(K+1)
01320        SMPA(K)=SMPA(K+1)
01330        DEXA(K)=DEXA(K+1)
01340        R(K)=R(K+1)
01350        R(K+1)=R(K+2)
01360        FLAG=FLAG+1
01370        IF(ERRORS.LT.10) THEN
01380        GOTO 20
01390        ENDIF
01400C
01410C       CALCULATION OF BIT ERROR RATE (B.E.R.)
01420C       --------------------------------------
01430        FLAG=FLAG-100
01440        PE(I)=REAL(ERRORS)/REAL(FLAG)
01450C
01460        PRINT 30,FLAG,ATB(J),SNR,DELTA,CMINUS
01465+       CZERO,CPLUS,PE(I)
01470    30  FORMAT(////,5X,'NO. OF PULSES=',I6/
01480+              5X,'ATB=',F4.0/
01490+              5X,'SNR=',F4.0/
01500+              5X,'DELTA=',F8.4/
01510+              5X,'CMINUS=',F8.4/
01520+              5X,'CZERO=',F8.4/
01530+              5X,'CPLUS=',F8.4/
01540+              5X,'PROBABILITY OF ERROR=',F12.8//)
01550        I=I+1
01560    40  CONTINUE
01570C
01580C       PLOTTING ROUTINE
01590C       ----------------
01600C
01610        PRINT 65
01620    65  FORMAT('PLOT REQUIRED (1=YES,0=NO)')
01630        READ *,N
01640        IF(N.EQ.0) THEN
```

Table 4.1
FORTRAN Simulation Program—Continued

```
01650          GOTO 85
01660          ENDIF
01670          I=1
01680C
01690C         PLOT VERTICAL AXIS
01700C         - - - - - - - - - - - - - - - - -
01710C
01720          PRINT 70
01730      70  FORMAT(20X,'PROBABILITY OF ERROR'/
01740+                5X,'-6',8X,'-5',8X,'-4',8X,'-3',8X,'-2'
01750+                8X,'-1'/3X,'10',5(8X,'10'),9X,'1'/
01760+                4X,'+',6('----------+'))
01770          DO 80 SNR=10,16,2
01780          PLOT=59+NINT(10*LOG10(PE(I)))
01790C
01800C         PLOT HORIZONTAL AXIS AND INDICATOR
01810C         - - - - - - - - - - - - - - - - - - - - - - - - - - - - - -
01820C
01830          PRINT 90,SNR,SPACE(1:PLOT)
01840      90  FORMAT(4X,'.'/4X,'.'/4X,'.'/
01850+                4X,'.'/4X,'.'/4X,'.'/
01860+                4X,'.'/4X,'.'/4X,'.'/
01870+                1X,F3.0,'+',A,'*')
01880          I=I+1
01890      80  CONTINUE
01900      45  CONTINUE
01910C
01920      85  PRINT 95
01930      95  FORMAT(//,'MORE VALUES? (1=YES,0=NO)')
01940          READ *,N
01950          IF(N.EQ.1) THEN
01960          GOTO 5
01970          ENDIF
01980          END
```

Table 4.2
BASIC Simulation Program

```
1  OPEN1,4
2  PRINT"ATB","SNR","COUNT","ERRORS"
3  PRINT#1,"ATB","SNR","COUNT","ERRORS,"
4  REM:
5  REM: INITIALIZE VARIABLES
6  REM:"————————————————————————"
7  ATB=1
8  FOR M=0 TO 2:ATB=ATB+M
9  FOR SNR=10 TO 16 STEP2
10 COUNT=0:ERRS=0
20 DELTA=0.10:K=1
30 PI=3.141592654
40 D(K)=1:Y(K)=0
50 R(K)=0:R(K+1)=0
60 CMINUS=0:CZERO=1:CPLUS=0
70 REM:
80 REM: RANDOM BINARY SEQUENCE GENERATOR
90 REM:"————————————————————————————————"
100 SIGSQ=10^(-SNR/10)
120 R1=RND(1):R2=RND(2):R3=RND(3)
130 D(K+1)=SGN(R1-.5+1E-38)
135 REM:
140 REM: RANDOM CHANNEL NOISE SIMULATION
145 REM:"————————————————————————————————"
150 VOLTSQ=-2*SIGSQ*LOG(R2)
160 NOISE=SQR(VOLTSQ)*COS(2*PI*R3)
165 REM:
170 REM: TRANSMIT AND CHANNEL FILTER
180 REM:"————————————————————————————"
180 Y(K+1)=D(K+1)*(1-EXP(-ATB/2))+Y(K)*EXP(-ATB)
190 R(K+2)=Y(K+1)+NOISE
195 REM:
200 REM: REGENERATOR SIMULATION
205 REM:"————————————————————————"
210 SAMPLD(K+1)=CMINUS*R(K+2)+CZERO*R(K+1)+CPLUS*R(K)
220 DHAT(K+1)=SGN(SAMPLD(K+1)+1E-38)
230 IF DHAT(K+1)<>D(K) THEN ERRS=ERRS+1
235 REM:
240 REM: 3-TAP ADAPTIVE EQUALIZER SIMULATION
245 REM:"————————————————————————————————————"
250 TERM=DHAT(K+1)-SAMPLD(K+1)
260 CMINUS=CMINUS+DELTA*TERM*R(K+2)
270 CZERO=CZERO+DELTA*TERM*R(K+1)
280 CPLUS=CPLUS+DELTA*TERM*R(K)
285 REM:
290 REM: REFRESH EQUALIZER VARIABLES
295 REM:"————————————————————————————"
300 D(K)=D(K+1)
310 Y(K)=Y(K+1)
```

Table 4.2
BASIC Simulation Program—Continued

```
320  SAMPLD(K)=SAMPLD(K+1)
330  DHAT(K)=DHAT(K+1)
340  R(K)=R(K+1)
350  R(K+1)=R(K+2)
360  COUNT=COUNT+1
365  REM:
370  REM: CHECK FOR END CONDITION
375  REM:"                              "
380  IF COUNT<10000 THEN 120
390  PRINTATB,SNR,COUNT,ERRS
400  PRINT#1,ATB,SNR,COUNT,ERRS
410  NEXTSNR
420  NEXTM
```

(1) *Random binary data sequence generator*. The programs generate a sequence of random binary symbols taking values $+1$ or -1 at random, each with probability 0.5. A binary code rather than ternary is chosen here for simplicity. It would not be difficult to modify the simulation to represent a ternary code, if required.

To generate the random sequence $\{d_k\}$, the simulation program causes a random number to be generated on each cycle of a logical loop. The random numbers lie in the range 0 to 1 with uniform probability distribution. Each time a random number n_k is generated, its value is checked against a "threshold value" of 0.5. If $n_k \geq 0.5$, d_k is taken to be $+1$. Otherwise d_k is -1.

(2) *Transmit and channel filter*. The programs simulate the transfer function of a low-pass filter to represent the channel. The basic pulse at the transmitter output is modelled as a rectangular pulse of width $T_b/2$.

The basic pulse $p_L(t)$ at the output of the channel is, therefore, of the form

$$
p_L(t) = \begin{cases} 0 & , \quad t \leq 0 \\ 1 - e^{-at} & , \quad 0 < t \leq T_b/2 \\ e^{-at}(e^{aT_b/2} - 1), & \quad t > T_b/2 . \end{cases} \tag{4.19}
$$

This is illustrated in Figure 4.16 for various values of aT_b (see also Exercise 3.1).

As indicated in the figure, the $p_L(t)$ of Equation (4.19) is equivalent to the output of a first order low-pass filter with 3dB bandwidth

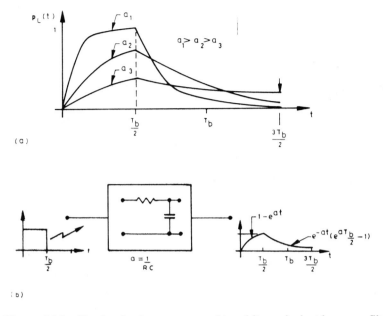

Figure 4.16   Simulated pulse response $p_L(t)$ and its equivalent low pass filter generation.

$$B = \frac{a}{2\pi} \quad (\text{Hz})$$

and with an input rectangular pulse of width $T_b/2$.

For an input sequence $\{d_k\}$, the channel output (regenerator equalizer input) for the simulated system is given by the pulse sequence

$$y_L(t) = \sum_{k=-\infty}^{\infty} d_k p_L(t - t_d - kT_b) + n(t).$$

From Figure 4.16, it can be seen that $p_L(t)$ has a trailing exponential "tail" which will cause ISI to subsequent pulses in the received data sequence. The function of the equalizer is to ensure that the ISI is minimized prior to the sampling and decision circuits. Consider the case where $d_o = 1$. At the channel output, the sampling instant for that pulse is at time $t = T_b/2$. Consider the instant $t = T_b/2$ in Figure 4.16. The subsequent pulses will be sampled at

$$t = 3T_b/2, \ 5T_b/2, \ \ldots .$$

As can be seen from the figure, the amount of ISI caused by $p_L(t)$ is a function of the constant a in Equation (4.19). For a narrowband channel (long-time constant), a is small. For a wideband channel, a is larger and the ISI is reduced.

For the equivalent low-pass channel in Figure 4.16(b), the bandwidth is $B = a/2\pi$. As a specific instance, a value of

$$aT_b = 2\pi$$

is equivalent to the channel having a bandwidth

$$B = 1/T_b.$$

The simulation allows various values of a to be used. This simulates variable channel bandwidth, and hence, varying amounts of ISI in transmission.

Let us now see how the computer programs simulate the $p_L(t)$ function given by Equation 4.19. Let y_L^k represent the kth sample of the channel output $y_L(t)$ to the regenerator in Figure 4.15. At this stage, let the additive random channel noise be ignored. Then commencing with $k=1$, the equalizer input samples will be

$$y_L^1 = d_1(1 - e^{-aT_b/2})$$

$$y_L^2 = d_2(1 - e^{-aT_b/2}) + d_1(1 - e^{-aT_b/2})e^{-aT_b}$$

$$y_L^3 = d_3(1 - e^{-aT_b/2}) + d_1(1 - e^{aT_b/2})e^{-2aT_b}$$

$$+ d_2(1 - e^{-aT_b/2})e^{-aT_b}.$$

The first term in each case is the component of the sample resulting from the "wanted" data input. Subsequent terms are the "unwanted" ISI components that result from previous data inputs.

Note that

$$y_L^3 = d_3(1 - e^{-aT_b/2}) + y_L^2 \cdot e^{-aT_b}.$$

In general, it can be seen that we can write (for the noiseless case)

$$y_L^{k+1} = d_{k+1}(1 - e^{-aT_b/2}) + y_L^k\, e^{-aT_b} \qquad (4.20)$$

where the first term represents the "wanted" data sample and the second represents the accumulated ISI that results from all previous pulses. We also need to prescribe the initial condition

$$y_L^1 = d_1(1 - e^{-aT_b/2}) \qquad (4.21)$$

and the simulation of the transmit and channel filtering is complete.

(3) *Random channel noise.* The simulation programs also provide for the representation of controlled amounts of Gaussian noise $n(t)$ being added to the channel output signal. The signal to noise ratio (SNR) at the regenerator equalizer input may be varied by the user.

The regenerator equalizer input signal is represented by discrete values y_L^k of the channel output. The signal component of y_L^k (in the noiseless case) is obtained using Equations (4.20) and (4.21). To represent additive Gaussian noise, we need to generate discrete noise values n_k with a Gaussian probability distribution, with mean zero and with specified variance σ^2. Then the $(k+1)$th regenerator input sample (signal plus noise) will be

$$y_L^{k+1} = d_{k+1}(1 - e^{-aT_b/2}) + y_L^k e^{-aT_b} + n_{k+1}. \tag{4.22}$$

Next we see how the value of the SNR can be set. Since the data input values d_{k+1} are either $+1$ or -1, then ideally for zero ISI ($aT_b \gg 1$) and zero noise, the channel signal output sample will be

$$y_L^{k+1} = d_{k+1}, d_{k+1} \in \{-1, 1\}$$
$$\text{for } aT_b \gg 1. \tag{4.23}$$

Since the signal value has magnitude 1 in this case, and since the noise has mean square voltage value σ^2, it is reasonable to define the *regenerator input SNR* in db as

$$\text{SNR (dB)} = 10 \log_{10}(1/\sigma^2). \tag{4.24}$$

Therefore, we can simulate different SNR values by setting the variance σ^2 of the noise samples to satisfy Equation (4.24).

At this stage an important problem remains. How do we program a computer to generate samples of a noise process such that the process is zero-mean Gaussian with a specified variance? Of course, most computer systems have random number generators which can be used to generate random independent numbers lying in the range 0 to 1. (We used such a random number generator earlier to generate our data sequence d_k.) The random number generators produce sequences of numbers that have equal probability of lying at any possible value between 0 and 1. That is, the numbers have a probability density function which is *uniformly* distributed from 0 to 1. We must find some way of transforming these random numbers so that they have a *Gaussian* probability density function. This is illustrated in Figure 4.17.

We can use the following procedure. Generate two uniform independent random numbers U_j and U_{j+1} in the range 0–1. Then a Gaussian random number X_j with mean μ and variance σ^2 can be obtained by computing

$$X_j = \mu + \sqrt{(-2 \sigma^2 \log_e U_j)} \cdot \cos(2\pi U_{j+1}), \quad j = 1, 3, 5, \ldots \tag{4.25}$$

For the proof of this relationship, see for example, Hartley (5).

(4) *Equalizer adjustment:* At each bit interval, the equalizer tap settings must be updated so that the equalizer behaves adaptively. In this simulation, the

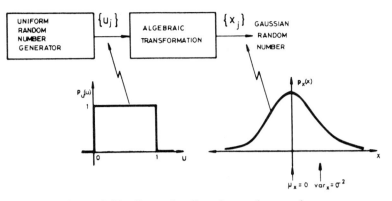

Figure 4.17 Generating Gaussian random numbers.

tap gain coefficients can be updated using Equation (4.18). That is, after each regenerator decision, the next tap value is set equal to the previous tap value plus an increment equal to

$$\Delta(\tilde{d}_k - y_k)r_n^{k+1}$$

where Δ is a positive incremental value to be chosen during simulation runs as discussed in Section 4.3.6 above.

(5) *Regenerator simulation:* In the simulation, the regenerator input is a series of samples consisting of components of the wanted signal, unwanted ISI, and unwanted noise. The sample values y_L^{k+1} are obtained from Equation (4.22).

At the equalizer and sampler output, the sample values are

$$y_k = \sum_{n=-N}^{n} C_n y_L^{k-n}. \tag{4.26}$$

These are the input values to the regenerator decision circuits. The regenerator decision threshold is at zero volts. Therefore, the simulation of the regenerator decision circuit simply consists of checking each y_k value and determining whether it is a positive or negative number. Then the regenerator output data estimates are given by

$$\tilde{d}_k = \begin{cases} +1 & , y_k > 0 \\ -1 & , y_k \leq 0 \end{cases} \tag{4.27}$$

The reader should now be in a position to work through the computer programs in Table 4.1 and/or Table 4.2 to see how the simulation proceeds. It might be preferable to write a new program to suit your own facilities. Whether you do so or not, you are strongly encouraged to run a simulation program (yours or one of those given) on a computer. Suggested parameter values for different runs are as follows:

(1) ISI

      Use aT_b = 1,2,4,6

(2) Noise

      Use   SNR (dB) = 10,12,14,16 (dB)

(3) Equalizer increment

      Use Δ = 0.05, 0.1, 0.2

Choose your own initial settings for the equalizer tap multipliers. Take note of how the multiplier C_n values are changed adaptively by the equalizer as the data sequence proceeds. Record BER values. Some typical results are shown in Figure 4.18. Note that the Fortran program in Table 4.1 computes and plots several points. It therefore takes considerable computation time. The Basic program in Table 4.2 computes only one BER value and therefore requires much less computation time. The latter approach does, therefore, enable the user to examine more rapidly the effects of parameter changes on the overall system performance.

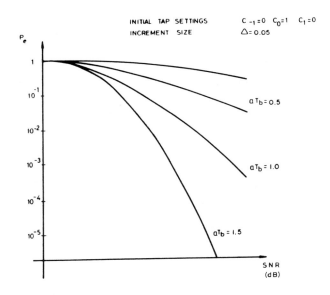

Figure 4.18 BER performance simulation results.

In the next section, we will examine methods for determining theoretical BER values for digital transmission systems. Analytical expressions for bit-error rates can be conveniently obtained for certain classes of relatively simple transmission system models. We will examine some of these. However, for many combinations of channel, type, interference phenomena, and equalizer-regenerator design, it is very difficult to find analytical forms for the expected BER. In these cases, computer simulation procedures such as the one above are very helpful.

4.4 BIT-ERROR RATE CALCULATIONS

In a baseband digital transmission system, there will always be a finite probability that transmission errors can occur. If the receiver equalizers are ideal, then the problem of ISI should be overcome. However, there still remain other sources of errors particularly

(1) *additive noise* including receiver and regenerator noise, crosstalk from other systems on pairs in the same cable and induced transmission line and equipment noise from other sources, and

(2) *timing jitter* resulting in reduced operating margins because of errors in locating regenerator sampling instants.

In this section we will examine techniques for theoretical estimation of bit-error rate (BER) values given certain operating conditions.

4.4.1 Mathematical Models

To analyze the BER performance of these systems, it is necessary to find suitable mathematical models for the system components and also for the noise and other source of error. So far as the transmitter, line, and regenerator are concerned, we have been using time invariant input-output functions in much of our earlier analysis. Referring to Figure 4.6, we have used transfer functions $P_T(f)$, $H_C(f)$ and $H_R(f)$ to represent the transmit filters, transmission line, and receiver equalizer filter, respectively. In doing this, we must bear in mind that in practical systems, these functions may not be time invariant. In particular, the transmission line characteristics, represented by $H_C(f)$, may vary considerably with ambient temperature variations and with aging. Therefore, our mathematical models are, in a sense, idealized models without which we could not perform the necessary analytical operations. It is sometimes difficult to ensure that calculated results will agree accurately with practical measurements for a wide range of operating conditions.

In the same way, we must use mathematical models to represent noise and other system impairments. Our primary concern is to calculate BER values expected for given transmit levels, system filtering, and noise conditions. In doing this, we must focus on the decision circuits in the regenerator and determine the (average) probability P_e that the decisions made will be in error. That is, we wish to determine

$$P_e = P_r\{\tilde{d}_k \neq d_k\}. \tag{4.28}$$

The regenerator decisions are based on the amplitudes of samples y_k of the received equalized waveform $y(t)$ as shown in Figure 4.6. The system schematic is shown in Figure 4.19 and the waveforms are illustrated in Figure 4.20. In this case, HDB3 coding is assumed.

The received equalizer output waveform may be represented

$$y(t) = \sum_k A_k p_r(t - t_d - kT_b) + n_o(t) \tag{4.29}$$

The noiseless component of $y(t)$ is illustrated in Figure 4.20(a).

Additive transmission noise represented by $n(t)$ in Figure 4.19 appears at the equalizer output as $n_o(t)$. This is illustrated in Figure 4.20(b). The sum $y(t)$ of this noise and the signal component is represented in Figure 4.20(c). The sample sequence $\{y_k\}$ is shown in Figure 4.20(d) together with the decision thresholds to be used by the regenerator decision circuits.

The resultant regenerator decoded output sequence $\{\tilde{d}_k\}$ is shown in Figure 4.20(e). It can be seen that an error occurred in the fourth symbol decision $(\tilde{d}_4 \neq d_4)$ because the sampled value y_4 fell below the upper threshold as a result of the addition of a negative noise spike at the sampling instant. (Note also that this single error is detectable after the eighth symbol has been regenerated because the sequence violates the rules for HDB3 coding discussed in Chapter 2.)

It should be clear that in computing the probability of error P_e for such a system, we will be concerned to characterize the noise process so that we can determine the probability that noise sample values will exceed certain limits. To do this, we naturally turn to probability density function descriptions of our noise processes.

It is often impossible to specify an accurate density function model to represent the additive noise signal $n(t)$. The most common model is to assume $n(t)$ has Gaussian (normal) characteristics. This is illustrated by a typical sample segment of a noise waveform $n(t)$ and its probability density function $p_N(n)$ in Figure 4.21.

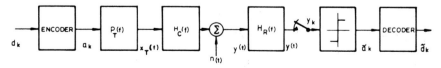

Figure 4.19 System schematic.

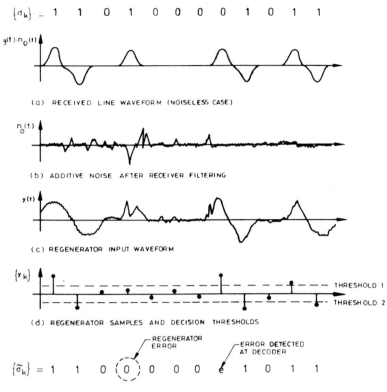

Figure 4.20 Waveform illustrating regenerator error as a result of noise.

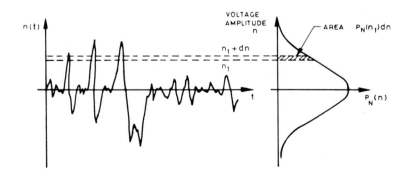

Figure 4.21 Representation of a probability density function for a Gaussian process.

Let a long sample of $n(t)$ be taken, and a measurement made of the function of time the waveform lies between the voltage values n_1, and $(n_1 + dn)$ volts where dn is a small increment. Then the measured fraction will be a good estimate of the value of the *probability density function* $p_N(n)$ at $n = n_1$ volts. In general, we write the probability that the noise (represented by N) will lie between voltage values n and $n + dn$ as

$$Pr\{n < N < n + dn\} = p_N(n)dn. \tag{4.30}$$

The function $p_N(n)$ described by the locus of points as n goes from $-\infty$ to $+\infty$ volts is called the probability density function. In the above simplified description, we have implied that the noise process is stationary, that is, its statistics do not vary with time.

For a *Gaussian* (or *normal*) noise signal we represent the probability density function as

$$P_N(n) = \frac{1}{\sigma\sqrt{2\pi}} e^{\frac{-(n - \mu_N)^2}{2\sigma^2}} \tag{4.31}$$

where μ_N represents the expected value (mean) and σ^2 represents the variance (note that σ is the rms value of the ac component of the noise). The Gaussian description of Equation (4.31) is the one most commonly used in calculations of BER performance in digital transmission. There are two main reasons for this:

(1) *Mathematical convenience*—The Gaussian noise model leads to the simplest analytical procedures, especially when the effects of filters have to be taken into account. If noise is Gaussian at the input of a linear filter, then it is also Gaussian at the output. (Its mean, variance, and power spectral density may of course be different.) This property is unique to the Gaussian probability density function.

(2) *Practical application*—Often the receiver input noise is the sum of a large number of components from different sources. It may be that the probability density of the individual source components is unknown or is not Gaussian (for example, it may be impulsive noise). Nevertheless, a remarkable phenomenon is represented by the *central limit theorem*, which indicates that a stationary random signal that is composed of a large number of independent random components, tends to have a Gaussian probability distribution. Several versions of this theorem have been proved in texts on statistics and verified experimentally. We will not discuss the details here.

A practical application is seen in the problem of crosstalk in PCM systems. If many PCM signals are being transmitted over adjacent pairs in the same cable, the crosstalk between pairs is often the dominant source of interference in repeater section design. In characterizing this crosstalk for design purposes, it has been found that, as predicted by the central limit theorem, a Gaussian distribution provides an accurate statistical model.

Sometimes, a given noise process is such that it is clearly not well represented by the Gaussian model. For example, impulsive noise may be induced in lines because of dialing impulses. This may be better represented by a probabilistic model which takes into account a higher probability of large spikes occurring. The *Laplace* probability density function is sometimes used in this case. It is represented by

$$p_N(n) = \frac{a}{2} e^{-a|n - \mu_n|} \tag{4.32}$$

which has mean μ_n and variance $\sigma^2 = 2/a^2$. This is a distribution in which the "tails" (high voltage values) fall off as e^{-n}, which is slower than the e^{-n^2} falloff for the Gaussian case. The Laplace probability density function is illustrated in Figure 4.22 for a zero mean random signal.

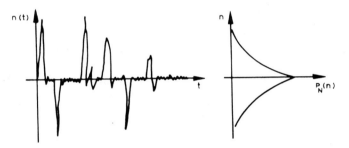

Figure 4.22   Laplacian random signal.

In the above description, we have referred to time averaging techniques for estimating the probability density function values for random waveforms. That is, the probability density function $p_N(n)$ is estimated in terms of the fraction of time the waveform lies between n volts and $(n + dn)$ volts as indicated in Equation (4.30). This is a valid procedure as long as the signal is ergodic (time averages equal ensemble averages). Time averaging procedures are usually more easily conceptualized by engineers.

However, as discussed in Chapter 2, statistical properties of random signals are usually defined in terms of ensembles of sample functions of the random signal. (The term "stochastic process" often appears, and is to the statistician, essentially the same as a "random signal" for an engineer.)

Figure 2.14 in Chapter 2 illustrates an ensemble or collection of all possible sample waveforms that the random signal source might generate. If we wish to analyze *sampled waveform* systems, then it is convenient to consider ensembles of samples of the discrete sampled waveform as shown in Figure 2.20 in Chapter 2.

The ensemble representation of a discrete random signal is important in understanding how we compute the performance of baseband digital transmission system. This is because of the fact that in most regenerator circuits of interest, the output regenerated bit sequence is determined on the basis of *samples* of the incoming signal. The regenerator samples the received equalized signal and compares the sample magnitude against one or more threshold values. In order to compute the probability of error, it is necessary to specify the statistics of the sample values.

If a random signal $x(t)$, is sampled at times

$$t = 0, T_b, 2T_b, \ldots,$$

then random variables

$$X_0, X_1, X_2, \ldots$$

can be visualized as the collection of possible sample value realizations at times $0, T_b, 2T_b, \ldots$. This was illustrated in Chapter 2, Figure 2.21.

If a continuous random process $x(t)$ is a Gaussian process, the random variables $X_0, X_1, X_2, \ldots$ associated with samples of the continuous process are also Gaussian.

If the process is strictly *stationary*, the discrete random variables $X_0, X_1, X_2, \ldots$ are identically distributed. That is, not only are they Gaussian but they have the same mean and variance. Furthermore, if the random signal is a *white* noise process, then the random variables are statistically independent as discussed in Chapter 2. Random variables that satisfy both of the above conditions are referred to as independent and identically distributed (iid).

Exercise 4.3

A white noise voltage $n(t)$ with amplitude 20 mV rms has zero dc component. Determine the probability that the value of a sample of the noise waveform will exceed $V_t = +0.1(V)$. Consider two cases, namely a Gaussian signal and a Laplacian signal, respectively.

Solution. Let the noise sample value be N volts. For the *Gaussian case*, the random variable N will have a probability density function given by Equation (4.31) with $\mu_n = 0$ and $\sigma^2 = 0.0004$ (V^2).

Then the probability that the amplitude N will be greater than 0.1 volt is given by the area under the probability density function for all values greater than V_t illustrated in Figure 4.23. This is given by the integral

$$P_r\{N > V_t\} = \int_{V_t}^{\infty} \frac{1}{\sigma\sqrt{2\pi}} e^{-n^2/2\sigma^2} \, dn.$$

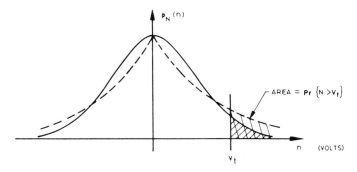

Figure 4.23 Threshold for random variable N.

It is not possible to solve this integral in closed form so numerical integration is used. Computed results are tabulated or plotted in terms of the function shown in Figure 4.24, namely

$$Q(x) = \frac{1}{\sqrt{2\pi}} \int_x^\infty e^{-t^2/2} \, dt. \tag{4.33}$$

To use this function to solve our problem we need to change the variable of integration. Let $t = n/\sigma$. Then we use

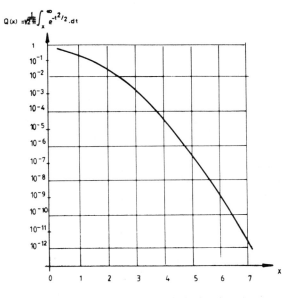

Figure 4.24 Area under Gaussian function.

$$dn = \sigma \, dt$$

$$\text{Lower limit} = V_t/\sigma$$

$$\text{Upper limit} = \infty$$

and so

$$Pr\{N > V_t\} = \int_{V_t/\sigma}^{\infty} \frac{1}{\sqrt{2\pi}} e^{-t^2/2} \, dt$$

$$= Q\,(V_t/\sigma)$$

and for $V_t = 0.1$ we obtain

$$P_r\{N > 0.1\} = Q\,(5.0).$$

Using Figure 4.24 we obtain

$$Pr\{N > 0.1\} = 3 \times 10^{-7}.$$

Next, we consider the *Laplacian noise case*. In this case

$$Pr\{N > V_t\} = \int_{V_t}^{\infty} \frac{1}{\sqrt{2}\sigma} e^{-n\sqrt{2}/\sigma} \, dn.$$

In this case the integral can be evaluated in closed form as

$$Pr\{N > V_t\} = -e^{-n\sqrt{2}/\sigma} \Big|_{V_t}^{\infty}$$

$$= e^{-V_t\sqrt{2}/\sigma}$$

and for $V_t = 0.1$ we obtain

$$Pr\{N > 0.1\} = 8.5 \times 10^{-4}.$$

4.4.2 Probability of Error–Wideband Gaussian Noise Case

Consider the simple model of a baseband transmission system shown in Figure 4.25. An input sequence $\{a_k\}$ is assumed to consist of independent three-level symbols. The channel is assumed wideband ($H_C(f) = 1$) with additive white Gaussian noise $n(t)$ of mean zero and variance σ_n^2. Then the channel output is

$$y_L(t) = x_T(t) + n(t).$$

Then the sample and decision circuit output is the estimate $\{\tilde{a}_k\}$ of the input sequence $\{a_k\}$. We wish to determine the probability of error which can be written

$$P_e = Pr\{\tilde{a}_n \neq a_n\}.$$

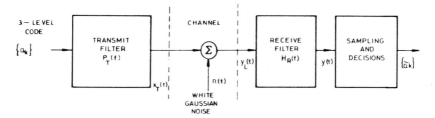

Figure 4.25 Model for BER analysis.

Wideband case: We consider the case where not only the channel is assumed wideband but so also is the receive filter. We assume that

$$H_R(f) = 1.$$

We also assume that the basic transmitted pulse shape $p_T(t)$ is a rectangular pulse of width T_b secs. That is, if $a_k = +1$, then a positive rectangular pulse of width T_b and amplitude V volts is sent. Likewise if $a_k = -1$, a negative rectangular pulse of width T_b and amplitude $-V$ is sent. If $a_k = 0$, no pulse is transmitted.

After the regenerator samples the received waveform, a resultant sample value will be

$$y_k = \begin{cases} V + N_k & , \quad \text{if } a_k = +1 \\ N_k & , \quad \text{if } a_k = 0 \\ -V + N_k & , \quad \text{if } a_k = -1 \end{cases} \tag{4.34}$$

where N_k is the kth sample value of the channel noise.

The regenerator decision circuit is assumed to make decisions based on two thresholds set at $+V/2$ and $-V/2$ volts, respectively. The output data estimate of a symbol $\tilde{a}_k$ is then made as follows

$$\tilde{a}_k = \begin{cases} 1 & \text{if} \qquad\qquad y_k > V/2 \\ 0 & \text{if} \quad -V/2 < y_k \leq V/2 \\ -1 & \text{if} \qquad\quad y_k \leq -V/2 . \end{cases} \tag{4.35}$$

To find the average bit-error rate P_e we need to take the average over the different possible values of transmit symbol a_k. We write this as

$$P_e = P_r(a_k = 1 \text{ and } y_k \leq V/2)$$

$$+ Pr(a_k = 0 \text{ and } y_k \leq -V/2 \text{ or } y_k > V/2$$

$$+ Pr(a_k = -1 \text{ and } y_k > -V/2) . \tag{4.36}$$

Consider the evaluation of the first term in Equation (4.36). Recall that for two events A and B we can write the joint probability

$$Pr(A \text{ and } B) = Pr(A|B)Pr(B) = Pr(B|A)Pr(A)$$

where $Pr(A|B)$ represents the conditional probability of event A occurring, given that B has occurred. Using this approach, we can write for Equation (4.36)

$$P_e = Pr(y_k \leq V/2 \mid a_k = 1) \, Pr(a_k = 1)$$

$$+ \, Pr(y_k \leq -V/2 \text{ or } y_k > V/2 \mid a_k = 0) \, Pr(a_k = 0)$$

$$+ \, Pr(y_k > -V/2 \mid a_k = -1) \, Pr(a_k = -1). \qquad (4.37)$$

Consider, for example, the last term in Equation (4.37). We can find

$$Pr(y_k > -V/2 \mid a_k = -1)$$

as follows. This is a similar problem to that of Exercise 4.3 in that we require the area under a probability density function for values of $y_k > -V/2$. Let the conditional probability density function for y_k (given that $a_k = -1$) be written $p(y|-1)$.

The result of adding a signal $+V$ to a zero mean Gaussian random variable is a Gaussian random variable with mean $+V$ and variance equal to the variance of the noise sample. From Equation (4.34) it follows that $p(y|-1)$ is Gaussian with mean $-V$ and variance σ_n^2. This is sometimes written in shorthand form

$$p(y|-1) = \eta(-V, \sigma_n^2).$$

Likewise, it is easy to see that y_k conditioned on $a_k = 0$ has density

$$p(y|0) = \eta(0, \sigma_n^2).$$

Also, y_k conditioned on $a_k = 1$ has density

$$p(y|1) = \eta(V, \sigma_n^2).$$

These probability density functions are illustrated in Figure 4.26.

Now we are in a position to evaluate Equation (4.37). Let us assume that

$$Pr(a_k = 1) = Pr(a_k = -1) = 0.25$$

$$Pr(a_k = 0) = 0.5.$$

Then Equation (4.37) becomes

$$P_e = 0.25 \int_{-\infty}^{V/2} \frac{1}{\sigma_n \sqrt{2\pi}} \, e^{\frac{-(y-V)^2}{2\sigma_n^2}} \, dy$$

$$+ \, 0.5 \left[1 - \int_{-V/2}^{V/2} \frac{1}{\sigma_n \sqrt{2\pi}} \, e^{\frac{-y^2}{2\sigma_n^2}} \, dy \right]$$

$$+ \ 0.25 \int_{-V/2}^{\infty} \frac{1}{\sigma_n \sqrt{2\pi}} \, e^{\frac{-(y+V)^2}{2\sigma_n^2}} \, dy.$$

By inspection of Figure 4.26, it is obvious that the first and last integral terms are equal and that the middle term inside the square brackets is twice that of the last integral. Therefore, we can write

$$P_e = [0.25 + 0.5 \times 2 + 0.25] \int_{-V/2}^{\infty} \frac{1}{\sigma_n \sqrt{2\pi}} \, e^{\frac{-(y+V)^2}{2\sigma_n^2}} \, dy.$$

As in Exercise 4.3, we change the variable of integration to

$$t = \frac{y+V}{\sigma_n}$$

so that

$$dy = \sigma_n \, dt$$

and the lower limit of the integral is

$$LL = \frac{V}{2\sigma_n} .$$

Hence, we obtain

$$P_e = 1.5 \ (Q(V/2\sigma_n) \tag{4.38}$$

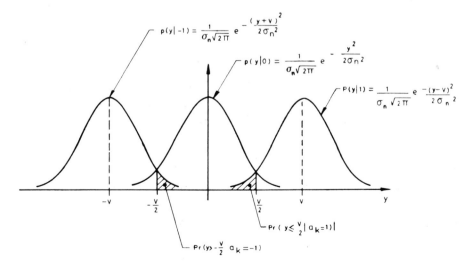

Figure 4.26 Conditional probability density functions for bit error rate calculation.

where the $Q(.)$ function is described by Equation (4.33). Note that P_e is a function of the ratio of signal amplitude V to the noise standard deviation σ_n. If we define a signal-to-noise power ratio term as

$$SNR = \frac{V^2/2}{\sigma_n^2} \qquad (4.39)$$

then we can write the bit-error rate as

$$P_e = 1.5 \, Q(\sqrt{SNR/2}). \qquad (4.40)$$

Exercise 4.4

Find the probability of error for the case where the signal to noise ratio defined in Equation (4.39) is equivalent to 10 dB.

Solution. Since

$$10 \log \frac{V^2}{2\sigma^2} = 10 \ (dB)$$

it follows that

$$P_e = 1.5 \, Q(\sqrt{5})$$

and using Figure 4.24 we obtain

$$P_e = 1.9 \times 10^{-2}.$$

Note that Figure 4.24 could be used directly as a plot of P_e versus SNR providing the scales are appropriately modified.

4.4.3   Allocation of Transmit and Receive Filtering

In the above error analysis, we assumed the channel and receive filters were wideband. This enabled us to compute P_e without too much complication. However, as we have seen, in most practical systems the signal is bandlimited by the channel and equalizer filter. Furthermore, the equalizer filter also performs a useful function in reducing the noise power at the input to the regenerator sampler and decision circuits.

In Equation (4.38) we saw that the probability of error in a regenerative repeater is a function of the peak signal to rms noise ratio (V/σ_n). If the noise variance σ_n^2 can be reduced, then P_e will be reduced.

In the above wideband analysis, the noise was assumed white, that is it has constant power spectral density. We can write this two-side spectral density as

$$S_n(f) = N_o/2 \quad (V^2/Hz).$$

This allowed us to use the fact that samples of the noise process were statistically independent.

Consider the system represented by Figure 4.25. As usual, we consider the sequence $\{a_k\}$ to be represented by a sequence of impulses (delta functions) with positive, zero, or negative amplitudes (weights). As we have seen, it is usually desirable that the receive signal $y(t)$ incorporate a basic pulse shape $p_r(t)$ which satisfies the zero ISI criteria specified in Equation (4.1).

In the wideband transmission system of Figure 4.25, the product of the transmit and receive filter transfer functions is

$$P_r(f) = P_T(f) H_R(f) \tag{4.41}$$

where $P_r(f)$ is the Fourier Transform of $p_r(t)$. The spectrum $P_r(f)$ must satisfy the Nyquist criterion. For example, $P_r(f)$ may be the raised cosine rolloff of Equation (3.12) with $\alpha = 0.5$.

The above prescribes the transfer function required of the transmit and receive filters in cascade. However, it does not tell us what is the best choice for each individual filter.

For one special class of assumptions, it is possible to show analytically how $P_T(f)$ and $H_R(f)$ could be allocated. This is for the case where the input symbols a_k are independent, the channel is assumed wideband, the noise white, and where we choose to allocate the pulse shaping so that the signal-to-noise ratio is maximized at the regenerator decision point. The noise is assumed only filtered by $H_R(f)$. The transmit filter output power P_t is assumed fixed at some maximum value allowed by such constraints as crosstalk or system loading.

The above assumptions are somewhat unrealistic. For example, if an HDB3 line code were used, the symbols a_k are not independent, as discussed in Chapter 2.

The transmit power can be obtained analytically by integrating the transmit signal power spectral density $S_T(f)$ over all values of f. That is, the transmit power is

$$P_t = \int_{-\infty}^{\infty} S_T(f)\, df. \tag{4.42}$$

Recall that in Chapter 2, we discussed how, for a given line code, we could obtain $S_T(f)$ using Equation (2.24) for the case of independent a_k symbols. As a result, the transmit power can be expressed in terms of the transmit filter response $P_T(f)$ as

$$P_t = \frac{E\{a_k^2\}}{T_b} \int_{-\infty}^{\infty} |P_T(f)|^2\, df. \tag{4.43}$$

P_t is set to the maximum value allowed by constraints such as crosstalk or system loading limitations.

The white channel noise $n(t)$ is filtered by the receive filter before reaching the regenerator decision point. Any linear operation (such as filtering) on a Gaussian process results in a Gaussian process. Therefore, the filtered noise $n_0(t)$ at the regenerator decision point is Gaussian and has power (in one ohm) of

$$\sigma_n^2 = \int_{-\infty}^{\infty} (N_o/2) \, |H_R(f)|^2 \, df \qquad (4.44)$$

where $H_R(f)$ is the receive filter response. We wish to choose the filtering $P_T(f)$ and $H_R(f)$ so as to minimize σ_n^2 given that P_t is set to a fixed (maximum) value.

To derive the optimum filtering allocation, it is useful to form the following inequality:

$$\int_{-\infty}^{\infty} \{|H_R(f)| - |P_T(f)|\}^2 \, df \geq 0.$$

This can be expanded and rearranged to give

$$\int_{-\infty}^{\infty} |H_R(f)|^2 \, df - 2 \int_{-\infty}^{\infty} |H_R(f)| \, |P_T(f)| \, df$$

$$+ \int_{-\infty}^{\infty} |P_T(f)|^2 \, df \geq 0. \qquad (4.45)$$

On substituting Equations (4.43) and (4.44) into (4.45) and using Equation (4.41) we obtain

$$\frac{2\sigma_n^2}{N_o} - 2 \int_{-\infty}^{\infty} |P_r(f)| \, df + P_t T_b/R_a(0) \geq 0.$$

We can write this

$$\sigma_n^2 \geq N_o \int_{-\infty}^{\infty} |P_r(f)| \, df - N_o \, P_T \, T_b/(2R_a(0)). \qquad (4.46)$$

Now we observe that for a given Nyquist spectral roll off, $P_r(f)$ is fixed. Also we assume P_t is fixed at its maximum value. Then the terms on the right-hand side of Equation (4.46) are constant independent of the choice of $P_T(f)$ or $H_R(f)$. In that case, the noise term σ_n^2 at the decision point must be a minimum when equality holds. Therefore, if we again substitute from Equations (4.43) and (4.44) into Equation (4.46), the noise term is minimized when

$$\int_{-\infty}^{\infty} |H_R(f)|^2 \, df = 2 \int_{-\infty}^{\infty} |P_r(f)| \, df - \int_{-\infty}^{\infty} |P_T(f)|^2 \, df. \qquad (4.47)$$

For this to be possible, it is necessary that the transmit and receive filters have transfer functions which satisfy

$$|P_T(f)| = |H_R(f)| = |P_r(f)|^{1/2}. \tag{4.48}$$

Hence, the regenerator noise power is minimized when the overall pulse shaping is equally divided between the transmitter and receiver. Such filters are sometimes referred to as "root-Nyquist filters."

Also, since $|P_r(f)|$ satisfies the Nyquist vestigal symmetry criteria, it can be shown that the autocorrelation function for the equalizer output noise has zeroes at the sampling instants. That is, the noise samples are uncorrelated. Since the noise is assumed Gaussian, it therefore follows that the samples are independent. Hence, the probability of error for each symbol is independent of the probability of error of its neighbors.

When Equation (4.48) is satisfied, it is possible to show that if a three-level code is used as described discussed in Section 4.4.2, then the average symbol error probability will be

$$P_e = 1.5 \, Q \left(\frac{3 \, P_T}{8 \, P_n} \right) \tag{4.49}$$

where P_T/P_n is the signal-to-noise power ratio at the input to the receiver filter and noise power is calculated in the Nyquist bandwidth $1/T_b$.

This root-Nyquist requirement for the transmit and receive filters is important in the design of some digital transmission systems such as those used in satellite and digital radio communications. However, it is seldom applied in line transmission systems where the assumptions used in the above derivation do not hold. Instead, in baseband digital transmission systems over cable pairs, the basic pulse transmitted to line is often just a rectangular pulse of width $T_b/2$. As discussed in Section 4.3, for that case the receiver filter equalizer must be designed in conjunction with the channel, to provide the pulse shaping that ensures minimum ISI.

4.5 PROBLEMS

4.1 Self-synchronization is the term often applied to the process of extracting clock timing information from the received random data signal.

This process cannot be achieved using a linear narrowband filter tuned to the clock frequency f_c unless the received data signal contains a discrete frequency component at $f = f_c$.

(1) Show graphically that a unipolar return-to-zero binary signal such as that shown in Figure 2.4 can be decomposed into two waveforms, the one a periodic waveform at the clock rate and the other a random

bipolar waveform. (Refer also to Problem 2.12.) As a result, a linear clock extraction method can be applied.

(2) Show that this is not possible if a unipolar nonreturn-to-zero signal is used or if a bipolar signal is used.

4.2 (1) Plot the voltage transfer function $|H_C(f)|$ for the cable pair whose insertion loss characteristic is assumed to be given at any frequency f by Equation (4.4). Assume $f_0 = 1024$ kHz. Find the 3dB bandwidth B.

(2) Compare your result in part (1) with the transfer function of an $R - C$ low-pass filter with the same bandwidth.

4.3 Repeat Question 4.2 for the case of a coaxial cable with loss given by Equation (4.5). Use $\alpha = 2.40$.

4.4 A 2048 kbit/s line transmission system uses half-width rectangular transmit pulses (see Exercise 4.1). The signal passes through a cable pair section with the transfer function obtained in Problem 4.2. It is then passed through an analogue equalizer filter.

Use graphical techniques to find approximately the equalizer's gain-frequency response $|H_R(f)|$ required to ensure the basic regenerator pulse shape $p_r(t)$ has a raised-cosine spectrum with $\alpha = 1$.

4.5 A single pulse input to a baseband transmission system results in received sample values for $p_L(k)$ of

$$0.0 \quad 0.0 \quad 0.2 \quad 1.0 \quad -0.2 \quad 0.1 \quad 0.0$$

for $k = -3, -2, -1, \ldots, 3$.

(1) Design a transversal equalizer that will eliminate ISI at the sample instants from $k = -2$ to $k = 2$ inclusive.

(2) Find the equalizer output $p_r(k)$ values for $k = -3$ to $+3$.

4.6 In computer simulation of digital transmission over a low-pass channel, Equation (4.20) represents the $(k+1)$th sample of the signal at the regenerator input for the case where half-width rectangular pulses are transmitted.

(1) Show how Equation (4.20) should be modified to model any transmission pulse width $T_w \leq T_b$.

(2) Repeat the simulation described in Section 4.3.7, allowing the pulse width T_w to be an additional variable and determine its effect on the bit-error rate at the equalizer/regenerator output.

4.7 A Gaussian noise voltage $n(t)$ has mean 1 (V) and variance 0.04 (V^2).

(1) Write an equation for its probability density function.

(2) Find the probability that the random noise voltage is less than 0.5 (V).

4.8 For baseband transmission over a wideband Laplacian noise channel, find an expression for the probability of error in terms of the received pulse amplitude V and the RMS value of the noise σ_n. Assume the use of ternary symbols with rectangular transmit pulses as in Section 4.4.2.

4.9 Use Equation (4.38) to determine the minimum signal-to-noise ratio necessary at the regenerator to ensure that a bit error rate of 10^{-5} or better is achieved.

4.10 Derive an expression similar to Equation (4.38) for the probability of error when binary unipolar transmit pulses of amplitude V are used on a wideband channel with additive Gaussian noise.

4.11 A binary bipolar baseband pulse with value $+V$ is transmitted over a line system. In the absence of noise, the regenerator input is affected by intersymbol interference. As a result, the sample value of the regenerator sampler input takes on one of three values. It is equal to V with probability 1/2. It is equal to $V+I$ or $V-I$ with probabilities 1/4, respectively, where I is a constant.

The system is now corrupted by additive Gaussian noise of mean zero and variance σ_n^2.

(1) Find an expression for the probability of error P_e in terms of V, I and σ_n^2.

(2) Plot curves of P_e versus SNR for values of $I = 0, 0.2V$ and $0.5V$, respectively.

4.12 The binary input to a noisy digital transmission system consists of 1's or 0's with probability $P(1) = P(0) = 0.5$. Figure 4.27 illustrates the transition probabilities on the channel, that is, the probabilities that a 1 will be received at the output of a regenerator given that a 1 was sent, that a 1 will be received given that a 0 was sent, and so on.

Find an expression for the probability of error P_e when $P(0|0) = P(1|1)$ and $P(0|1) = P(1|0)$, that is, the model represents a binary symmetric channel.

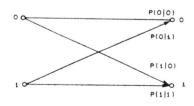

Figure 4.27

4.6 REFERENCES

1. R. W. Lucky, J. Salz and E. J. Weldon, Jr., *Principles of Data Communication*, McGraw-Hill, 1968.

2. N Q. Duc, "Transmission Rates and Error Control in High Speed Digital Coaxial Systems," Aust. Post Office Research Labs., Report No. 6834, Dec. 1973.

3. A. D. Whalen, *Detection of Signals in Noise*, Academic Press, 1971.

4. J. G. Proakis, "Advances in Equalization for Intersymbol Interference," from *Advances in Communication Systems*, Vol. 4, A. J. Viterbi (Ed.), Academic Press, 1975.

5. M. G. Hartley, *Digital Simulation Methods*, Peter Peregrinus Ltd., 1975.

6. K. S. Shanmugan, *Digital and Analogue Communication Systems*, Wiley, 1979.

7. P. B. Bylanski and D. G. W. Ingram, *Digital Transmission Systems*, Peter Peregrinus, 1980.

8. W. C. Lindsey and M. K. Simion, *Telecommunication Systems Engineering*, Prentice-Hall, 1973.

9. J. J. Spilker Jr., *Digital Communications by Satellite*, Prentice-Hall, 1977.

10. J. J. Stiffler, *Theory of Synchronous Communications*, Prentice-Hall, 1971.

11. CCITT: *Digital Networks—Transmission Systems and Multiplexing Equipment*, Recommendations G. 700—G. 956, Red Book, Vol. III, Geneva 1985.

Chapter 5

MEASUREMENT TECHNIQUES

5.1 INTRODUCTION

With the introduction of digital transmission into telecommunications networks, appropriate performance measurement techniques must be devised. In this chapter, we examine some of the principal testing procedures used in the laboratory and in the field to characterize transmission performance in digital systems.

Digital transmission facilities in telephony networks can be conveniently divided into two levels, namely inter-exchange PCM links, and subscriber-level digital communications. Transmission error measures are important in characterizing the performance at both these levels.

In Chapters 2–4, we examined transmitter and regenerator techniques for digital line systems. The performance of digital terminal and regenerator equipment is evaluated by measuring such factors as:

(1) error rates
(2) noise margins
(3) regenerator timing jitter

When error patterns occur, they may occur in a random fashion. In that case, the error events are statistically independent. This is characteristic of error patterns in certain digital transmission systems such as satellite circuits. However, for most practical digital line systems, errors tend to occur in patterns containing a mixture of random and burst errors. This is because of the nature of the crosstalk and interference that causes the errors. Errors can result from noise (including crosstalk) and from intersymbol interference resulting from pulse dispersion.

A number of different procedures have been developed to provide meaningful measures of the transmission errors and their effect on circuit availability and quality. These measures include average bit-error rates (BER), average block-error rates, and error-free seconds (EFS). The measurement of error rates will be discussed in detail in this chapter. We will also examine the properties of pseudo-random binary test signals commonly used in many of these measurements.

In the laboratory setting, a useful way for evaluating the performance of filters, equalizers, and timing extraction circuits is by means of *eye-diagrams*. They also provide useful insight into the parameters effecting transmission error rates. We will first examine these eye-diagram measurement techniques.

5.2 EYE-DIAGRAMS

5.2.1 Measurement Procedure

An eye-diagram is a synchronized display on an oscilloscope of superimposed segments of a digital line waveform. Each waveform segment is usually $2T_b$ (sec) long where T_b is the bit interval. Figure 5.1 shows a block diagram of the measurement procedure.

A pseudorandom binary sequence (PRBS) generator is used to generate a pattern of random binary symbols. This is sometimes called a pseudo-noise (PN) sequence. The sequence is noiselike in that it approximates a random sequence such as would be generated by a sequence of coin tossings in which a 1 is represented by a head and a 0 is represented by a tail. It is called pseudorandom because it is not truly random but is generated by a feedback shift register circuit and is, therefore, deterministic. However, it has many properties in common with a truly random binary sequence. Pseudorandom binary sequences will be discussed in much greater detail later in this chapter.

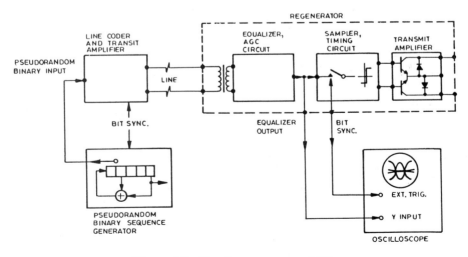

Figure 5.1 Eye-diagram measurement.

In the eye-diagram measurement schematic of Figure 5.1, the output of the pseudorandom binary sequence generator is passed into the line coder and digital transmission system. After transmission through the digital line system, the voltage waveform $y(t)$ at the input to the regenerator sampler circuit is connected to the Y amplifier of an oscilloscope.

A bit rate synchronizing signal is connected to the oscilloscope external trigger input. This may be derived from the bit timing extraction circuit at the regenerator. Alternatively, the bit synchronizing signal may be obtained directly frrom the PRBS generator if the test configuration involves having the PRBS generator in the same location as the oscilloscope.

The time base of the oscilloscope is set to trigger at the center of a bit interval with a sweep duration of two bit intervals. The resulting eye-diagram is displayed on the oscilloscope as the superimposition of all possible received waveform transitions of length $2T_b$ commencing at a symbol sampling instant and with all sampling instants lying at the same points.

Exercise 5.1

The binary sequence

$$\{b_k\} = \{\,1\ 0\ 1\ 1\ 0\ 1\ 0\ 1\ 0\ 0\ 1\ 0\ 1\ 1\ 0\ 0\,\}$$

is transmitted using an alternate-mark-inversion (A.M.I.) line code. The resultant waveform $y(t)$ at the sampler input of the regenerator is given by

$$y(t) = \sum_{k=0,1,\ldots} a_k\, p_r(t - kT_b)$$

where the line code symbol sequence is

$$\{a_k\} = \{+1\ 0\ -1\ +1\ 0\ -1\ 0\ +1\ 0\ 0\ -1\ 0\ +1\ -1\ 0\ 0\}.$$

The basic receive pulse shape is raised cosine, that is

$$p_r(t) = 0.5 + 0.5\ \cos \frac{(\pi t)}{2T_b}, \qquad -T_b < t \le T_b.$$

This is shown in Figure 5.2. Sketch the eye-pattern generated by the waveform $y(t)$.

Solution. The sampler input waveform $y(t)$ is drawn in Figure 5.3(a). Then each successive $2T_b$ (sec) segment of $y(t)$ is superimposed in turn on the initial segment to obtain the eye-pattern shown in Figure 5.3(b). Note that not all possible a_k transitions are allowed by the A.M.I. coding. The symbol pairs $(1, 1)$ and $(-1, -1)$ cannot occur in the sequence.

Corresponding points A, B, ... , G are shown on the $y(t)$ waveform and eye-diagram, respectively.

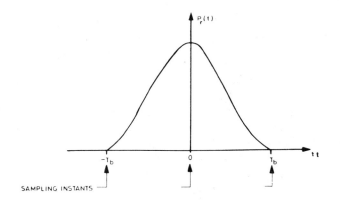

Figure 5.2   Basic receive pulse shape $p_r(t)$.

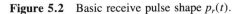
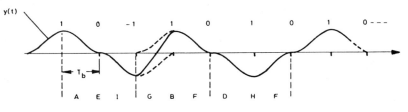

(a) SAMPLER INPUT WAVEFORM

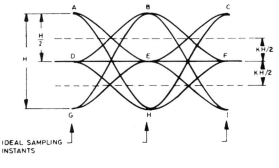

(b) EYE-DIAGRAM

Figure 5.3   Evolution of an eye-diagram for a ternary line waveform.

5.2.2   Important Features of Eye-Patterns

The pattern in Figure 5.3(b) represents a typical eye-pattern for a noise-free ternary line signal with a raised-cosine basic pulse shape. The main feature of the pattern that gives it its name is the pattern of eye shapes. The eyes result from the separation of the received signal levels into one of three voltage values at the sampling times. The eye closes between sampling points because of transitions between symbol values.

Figure 5.4 shows typical eye-patterns for a two-level digital signal. (Although ternary line codes are most popular in many digital line systems, binary codes are often used in optical fiber systems, subscriber lines, and some radio systems.) Figure 5.4(a) represents an idealized noise-free binary signal with no distortion. The effects of noise and distortion are illustrated in Figure 5.4(b).

Certain features of an eye-pattern are useful indicators of error rate performance. They are illustrated in Figure 5.5. The key features are:

(1) *eye opening*—the minimum separation at the sampling times between the two signals representing different adjacent symbol values.

(2) *noise margin*—another term used to quantify the eye opening. It is a measure of the shortest distance from an eye boundary (at the sampling time) to the midpoint between ideal received signal levels.

(3) *eye width*—the minimum width of the eye at the midpoint between ideal signal levels

(4) *rate of closing of the eye*—the slope of the eye boundary in the vicinity of the optimum sampling time

(5) *zero crossing distortion*—the amount of distortion in the vicinity of the zero crossings (width of the eye boundaries at the zero crossings).

The *noise margin* or eye opening indicates the susceptibility of the regenerator to crosstalk and interference. It can be used to provide an estimate of the regenerator error rate. In general, the regenerator decision threshold will be set at or near the voltage value midway between ideal received signal levels. Then the noise margin represents the smallest amount of additional noise which, if added to a received signal, would cause an error. An increase in the noise margin will normally decrease the error rate. The noise margin or eye opening indicate the

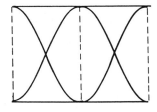 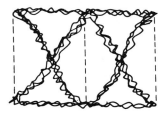

(a) IDEAL NOISE-FREE CASE   (b) DISTORTION AND NOISE EFFECTS

Figure 5.4 Eye-patterns for a binary signal.

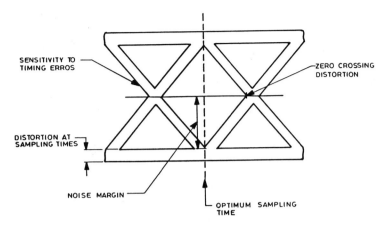

Figure 5.5 Eye-pattern features.

effectiveness of the design of the equalizer. A wide open eye indicates that the amplitude and phase distortions introduced by the channel have been minimized. In practice, a given equalizer circuit will usually be required to cope with a range of transmission line lengths. The effectiveness of the design of an equalizer in a regenerative repeater is indicated by its ability to maintain high noise margins as the line length of the repeater section is varied.

While the noise margin or height of the eye indicates the margin against signal amplitude disturbances, the *eye width* indicates the regenerator margin for timing variations. The optimum sampling instants are located where the eye opening is a maximum. Variations of the sampling instants from the optimum will reduce the noise margin and, as the edges of the eye are approached, the error rate will increase rapidly. These effects are a function of the eye width. If the eye width is small, then there is a need for careful design of the bit synchronization recovery circuits if the error rate is to be kept low.

The sensitivity of the regnerator system to timing errors is also affected by the *rate of closing of the eye* (the slope of the eye in the vicinity of the ideal sampling point). If this slope is large, then small timing variations lead to large reductions in noise margin. If the rate of closure is low, small timing variations have negligible effect on the bit error rate.

The *zero crossing distortion* is the variation in the zero crossings of the eye. The zero crossings correspond to the time instants at which the eye waveforms pass through zero volts. The zero crossing distortion is an indication of the likely extent of variations at the output of the timing recovery circuits which set the sampling instants. Ideally, if adjacent symbols have different values either side of zero, the transition waveform should cross through zero at an equal distance from the sampling time for each symbol.

A *timing offset* is said to occur if the zero crossing transitions consistently occur earlier or later than the midpoint between ideal sampling instants. Variations in the sampling point that result in approximately equal numbers of early and late samples are called *timing jitter*. Although timing offset, if detected, can be removed by shifting the sampling time, timing jitter is always likely to be present to some extent.

5.2.3 Effects of Intersymbol Interference

For a ternary signal (such as A.M.I. coding), the eye-diagram shown in Figure 5.3(b) is typical of that obtained when no inter-symbol interference (ISI) or noise is present. Note how, at the sampling times, all traces pass exactly through one of the three symbol values $+H/2$, 0, or $-H/2$. Furthermore, all the pulse "tails" pass through 0 volts at the sampling instants.

The regenerator output decisions are based on a comparison of the magnitude of each sample value with two threshold levels $+KH/2$ and $-KH/2$, as discussed in Chapter 4. As mentioned above, it can be shown that, in practice, the optimum threshold levels for the regenerator decision circuits are at approximately half the height of the eye, that is, where $K = 1$.

Figure 5.6 shows an eye-diagram in which the effects of ISI are evident. This diagram was obtained in a practical experiment when a loop was connected directly from a transmitter output to regenerator input at a terminal. The regenerator equalizer was designed to operate satisfactorily for lines with loss values in the range 5 to 35 dB. (Loss is specified at the frequency equal to one half the bit rate.) In this experiment, some ISI is evident when the loss is zero.

A careful examination of Figure 5.6 indicates that a transition from the $+1$ symbol to 0 results in undershoot at the 0 sampling instant. As a result, the sample will have a negative value where the regenerator voltage should be zero. Likewise a -1 to 0 transition results in overshoot. These two effects result in a distortion band of h_2 volts in the 0 symbol value at the sampling instant.

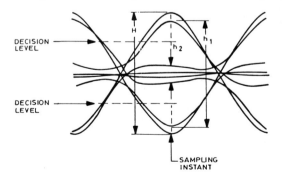

Figure 5.6 A practical eye-pattern with ISI evident.

Next, we consider the distortion at sampling instants at the top and bottom of the pattern shown. For the $-1, 1$ transitions and $1, -1$ transitions, the equalizer output is approximately of raised cosine shape with height H. On the other hand, a difference of h_1 volts separates the ends of the $0, 1$ and $0, -1$ transitions at the sampling instant, with $h_1 < H$. That is, the value of a sample associated with a transmitted 1 may be either $+H/2$ or $+h_1/2$ volts depending on whether the previous symbol was -1 or 0, respectively. This is a clear illustration of ISI, since the sample values at one sampling instant are dependent on symbol values at previous instants.

We can express the degradation in noise margin as a result of these ISI effects as follows. Assume the regenerator decision threshold levels are set at $+KH/2$ and $-KH/2$, respectively. For 0 symbols, the noise margin has been reduced from $KH/2$ with no ISI to $(KH/2 - h_2/2)$ when ISI is present. Hence, we obtain

$$\text{``0'' Noise margin degradation} = 20 \log_{10} \frac{KH/2}{KH/2 - h_2/2} \qquad (5.1)$$

for transmission of 0 symbols.

Exercise 5.2

Find the degradation in noise margin resulting from the ISI for the cases where 1 and -1 symbols are transmitted.

Solution. It is straightforward to show that

$$\text{``1'' Noise margin degradation} = 20 \log_{10} \frac{(1 - K)H/2}{(h_1/2 - KH/2)} \qquad (5.2)$$

for either 1 or -1 transmitted.

As an indication of typical numerical values involved in the above noise margin degradations, consider the case where:

$$H = 5 \text{ volts, } h_1 = 4.5 \text{ volts, and } h_2 = 0.6 \text{ volts.}$$

If we assume $K = 0.5$ then we obtain

$$\text{Noise margin degradation} = \begin{cases} 2.4 \text{ (dB) for 0 symbols} \\ 1.9 \text{ (dB) for 1 or } -1. \end{cases}$$

As mentioned previously, it is usual in practical regenerator systems for the noise margin degradation to be a function of line length. Equipment designers aim for equalizer characteristics which can provide good performance over a

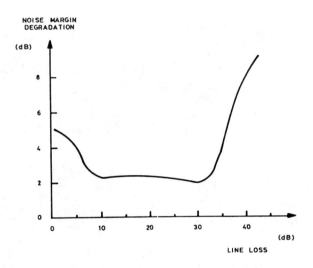

Figure 5.7   Typical regenerator noise margin degradation with line loss.

range of line loss values (for example 5 to 35 dB). Figure 5.7 illustrates the typical performance of a 2 Mbit/s regenerator operating with various cable attenuation values in the absence of crosstalk. Further details are given in Semple (1).

5.2.4   Effects of Noise and Crosstalk

In practical transmission systems, the eye-diagram at the regenerator equalizer output will be affected by noise and crosstalk as well as ISI. Figure 5.8(a) shows a typical eye-diagram with random noise and ISI present. Figure 5.8(b) shows the additional effect of crosstalk from another digital system operating at the same bit rate in an adjacent cable pair.

   Crosstalk from other digital sources can not only reduce the noise margin but also have a serious effect on timing jitter. When digital systems are operating over a common cable, they are often controlled by independent, but nominally identical clock signals. As a result, the effects of such crosstalk on the eye-pattern is such that an interference pattern moves relatively slowly through the diagram in the direction of the time axis, at a rate proportional to the small differences in clock frequencies. For more details concerning the effects of crosstalk on timing jitter, see for example Feher (2).

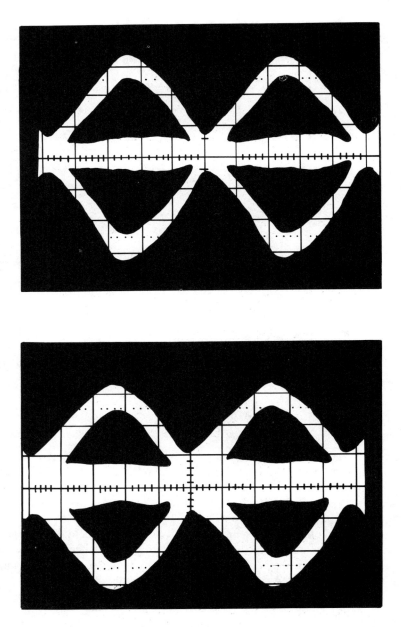

Figure 5.8   Eye-diagrams with ISI, noise, and crosstalk.

5.3 NEAR-END CROSSTALK NOISE FIGURE

5.3.1 Regenerator Performance Measurement

The error rate performance of many digital transmission systems is primarily limited by the effects of crosstalk. For example, in 2 Mbit/s and 1.544 Mbit/s junction PCM circuits, near-end crosstalk (NEXT) is often the dominant source of interference which limits the spacing between regenerators. For subscribers' 80 kbit/s and 144 kbit/s 2-wire burst systems to be discussed in Chapter 6, Volume 2, far-end crosstalk (FEXT) may be more important. Crosstalk into a regenerator is classified as NEXT or FEXT depending on whether the interference sources are local or at the remote end of the repeater section, respectively.

The performance of a regenerator operating in a crosstalk limited environment can be assessed in a more quantitative manner than by examination of eye-diagrams. It is possible to measure a *NEXT noise figure* and a *FEXT noise figure* for the regenerator which may be more suitable for performance specification. These parameters measure the amount of crosstalk required to produce a specified error rate. The crosstalk is simulated by a wideband noise signal generator. The noise signal is filtered to produce a test signal with power spectral density characteristics representative of typical crosstalk types.

Figure 5.9 illustrates the principles involved in measuring the effects of NEXT on regenerator performance. The noise source is intended to represent a large number N of PCM regenerator transmitters. For this purpose, it is desirable that the noise source is Gaussian with power spectral density $S_T(f)$ equal to the average transmitter output power spectral density. For instance, $S_T(f)$ may represent the spectrum of half-width rectangular basic pulse shapes with HDB3 line coding as discussed in Chapter 2, Volume 1.

The coupling of the NEXT effects from the N transmitters to the regenerator under test is simulated by a filter with transfer function $H_x(f)$ and an adjustable attenuator with loss $[R_N]$. Note that quantities in square brackets are taken to be

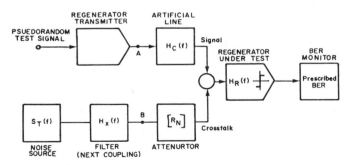

Figure 5.9 Measurement of NEXT noise figure.

in decibels in this section. The filter transfer function simulates the crosstalk coupling mechanism. In practice, for balanced cable pairs, this has been found to be equivalent to a coupling circuit with an amplitude response which falls off at 15 dB/decade and can be approximated by

$$H_x(f) = K|f/f_o|^{3/4} \tag{5.3}$$

where f_o is equal to half the system bit rate.

At the output of the attenuator, the resultant "crosstalk" is added to the desired "signal." The latter is obtained by connecting a pseudorandom source, to a transmitter and artificial line simulator. The line may be simulated by an appropriate linear network. For example, a cable characteristic with loss represented by Equation (4.3) could be assumed. Typically, the network is designed to simulate a cable with 30dB loss at f_o.

Then the rms voltages at points A and B are adjusted to be equal by varying the level of the noise source. On completion of this calibration procedure, the attenuator loss value $[R_N]$ is adjusted so that the crosstalk noise produces a specified bit error rate (say 10^{-7}) at the BER monitor.

Let N_o be the mean-square NEXT interference at the regenerator decision point producing the specified error rate of 10^{-7}. Also, we define a regenerator noise amplification factor

$$I_N = \int_{-\infty}^{\infty} S_T(f)|f/f_o|^{3/2}H_R(f)df. \tag{5.4}$$

This reflects the amount of crosstalk noise produced by the regenerator equalizer $H_R(f)$.

Then the NEXT noise figure of the regenerator under test is defined in dB as

$$[R_N] = [I_N] - [N_o]. \tag{5.5}$$

The noise figure $[R_N]$ is a measure of the quality of the regenerator design. The $[I_N]$ factor can be reduced by reducing the bandwidth of the equalizer. The $[N_o]$ factor depends on parameters which affect the decision process such as

(1) impairments from ISI
(2) offsets from optimum of the decision threshold levels
(3) sampler timing errors

If the bandwidth of the equalizer is reduced too far, it leads to an increase in ISI, which reduces $[N_o]$ for a given bit error rate. In any regenerator design, there is therefore a tradeoff between $[I_N]$ and $[N_o]$.

The measurement of the NEXT noise figure $[R_N]$ permits comparison of the performance of different regenerators. In theory at least, the immunity of PCM regenerators to NEXT could be calculated from a knowledge of the regenerator equalizer characteristics, the resultant ISI statistics, and circuit design parameters involving the clock extraction process and decision level control. In practice, this calculation of $[R_N]$ would be complex. In any case, a theoretical value might be of limited validity because of circuit realization limitations. Therefore, the NEXT noise figure is used as a convenient regenerator performance measure to be determined by physical measurement.

5.3.2 Input Signal to NEXT Noise Ratio

An alternative performance measure is the *input signal* to *NEXT noise ratio*. This is defined as the ratio of the signal level at the artificial line output in Figure 5.9 to the crosstalk noise level which when added to the signal, produces a specified bit error rate. A number of test instruments are commercially available to measure input signal to NEXT noise ratio. These are useful for assessing the relative performance of regenerators with the same transmit signal power spectral densities $S_T(f)$.

In general, however, the NEXT noise figure has been shown to be a superior measure with measurement errors of less than 0.5 dB using practical noise-shaping filters. More details are given in Semple and Gibbs (3 and 13).

5.4 PSEUDORANDOM BINARY TEST SIGNALS

5.4.1 Introduction

In the display of eye-diagrams and in the measurement of error rates, it is necessary to use a digital test signal that has random characteristics. In this section, we consider methods of generating such signals.

The amount of intersymbol interference present at any instant at the regenerator is usually a function of the particular symbol pattern that occurred prior to that instant. Recall that in the previous section, it was noted that the waveform segment shapes, which made up the eye-pattern, were functions of the sequence of symbols involved. Where intersymbol interference results in regenerator errors, some particular transmitted symbol sequences will give rise to more errors than others.

It is, therefore, desirable in performance measurements, to use test signals which are designed to give rise to all possible symbol sequence patterns. Ideally, a purely random binary input to the transmission system would be desirable.

Pseudorandom binary sequence signals are a convenient means of approximating to a purely random binary signal. Their primary advantage is that they are easily generated by feedback shift registers. Because they are important in many measurement applications, we will examine the methods used for their generation and their properties.

5.4.2 Searching for a Random Sequence

Before examining the circuit techniques used for generating pseudorandom sequences, it is instructive to ask what are the essential characteristics required of the sequences to be produced. In some sense we want the sequence to appear "as random as possible." But what is our test for randomness?

Imagine, for example, that we are conducting a young designers competition in which the designers are required to find a source which generates binary sequences which are "as random as possible." Unfortunately, truly random sources are not easy to come by, so we hypothetically offer a valuable prize for the best circuit design. Three "black boxes" have been entered in the competition and it is now up to us to decide which is to be awarded the prize. How do we proceed to make the decision?

A reasonable approach would seem to be to simply turn each of the black boxes on, allow them to generate a sample sequence, and then to examine the results. Let us say that, in doing so, we obtain the following output sequences:

From box A: 1 1 1 0 1 0 1 1 0 0 0 0 1 0 1 0

From box B: 1 0 1 0 1 0 1 0 1 0 1 0 1 0 1 0

From box C: 0 1 1 0 1 0 0 1 0 1 1 0 1 0 0 1

Now what tests for randomness shall we apply? We might be tempted to immediately discard box B because it seems to be producing a periodic sequence but not a random one. However, the designer of box B protests. He argues that if the sources were truly random, then all three sequences are equally likely. He is of course quite correct. But somehow we suspect box B is just a square wave generator.

It seems that if we examine a sequence after it has been produced, we may never conclude whether or not the source is a "truly random one" unless we have some *a priori* knowledge about the generator method. We are, therefore, forced to prescribe a number of features that we expect a random sequence to exhibit. A sequence that passes these tests is called *pseudorandom*. The sequence must have the attributes of a random sequence even though it may have been generated by a nonrandom deterministic source.

Exercise 5.3

Devise suitable tests to decide whether to choose box A, B, or C for the award of the prize.

Solution. The sequence ought to be such that any statistical test on it gives results which are indistinguishable from those resulting from the same test applied to a sample sequence from a truly random source of independent symbols. Such tests might include:

(a) Mean values

$$Pr(0) = Pr(1) = 0.5?$$

(b) Independence

$$Pr(0|0) = Pr(0) = Pr(0|1)?$$

$$Pr(1|0) = Pr(1) = Pr(1|1)?$$

Note that all boxes pass the first test equally. Box B fails the second miserably! The sequence from Box C suggests periodic repetition after eight bits. We really need longer sample sequences to be sure, but we award the prize to A anyway!

5.4.3 Feedback Shift Register Generators

A chain of shift registers with appropriate feedback connections is a convenient way of generating a pseudorandom sequence. The sequences are actually deterministic and periodic. However, they will satisfy any tests for randomness over segments of time shorter than one period.

Figure 5.10 shows a general feedback shift register structure. Note that waveform changes between the two binary signal levels can only occur at regularly spaced intervals Δt determined by an input clock of frequency f_c where

$$\Delta t = 1/f_c.$$

Initially, the shift registers are set to some initial state other than the all-zeroes state. Then the clock signal successively shifts the contents of the shift-register

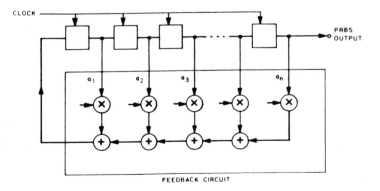

Figure 5.10 Typical PRBS generator structure.

stages to the right. At the same time, a sequence of feedback bits is fed back to the input of the first stage.

The output binary sequence values are determined by the structure of the feedback network. A set of simultaneous equations describes the new contents x_1^i of the ith stage in terms of the old contents x^i prior to each clock pulse.

For an n-bit shift register, the equations are

$$x_1^n = x^{n-1}$$

$$x_1^{n-1} = x^{n-2}$$

$$. . .$$

$$x_1^2 = x^1, \text{ and}$$

$$x_1^1 = a_1x^1 + a_2x^2 + . . . + a_nx^n. \tag{5.6}$$

The feedback digit is represented by the last of the equations which is referred to as the feedback equation. The addition is carried out modulo-2 (exclusive-OR operation). The coefficients a_i for $i = 1$ to $n-1$, are either 1 or 0, representing the presence or absence, respectively, of a connection from shift register element i to the feedback structure.

For an n-bit feedback shift register, there are 2^n different possible feedback connections. Each will give rise to a different output sequence. In practice $a_n = 1$. Otherwise, we would have the degenerate case where the last stage plays no part in the sequence generation.

There are, therefore, 2^{n-1} possible feedback structures for a given n-bit shift register. Next, we will examine how the choice of feedback structure influences the output binary sequence. Consider the two different examples of 4-bit feedback shift register structures shown in Figures 5.11(a) and (b). The feedback equations for these examples are

$$x_1^1 = x^2 + x^4 \text{ for case (a), and}$$

$$x_1^1 = x^1 + x^4 \text{ for case (b).}$$

Exercise 5.4

If the initial contents of the shift registers in Figures 5.11(a) and (b) are set to 1111, determine the set of shift register contents after each successive clock cycle.

Solution. The register contents for Figure 5.11(a) are

	CYCLE			CYCLE
1111, 0111, 0011, 1001, 1100, 1110,			1111, 0111, . . .

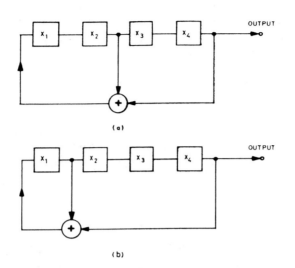

Figure 5.11 Feedback shift register examples for $n = 4$.

Note that the output sequence is

$$1\ 1\ 1\ 1\ 0\ 0\ 1\ 1\ \ldots$$

which is periodic with cycle length 6. On the other hand, the shift register contents for Figure 5.11(b) are

|_____CYCLE_____
¦ 1111, 0111, 1011, 0101, 1010, 1101, 0110, 0011, 1001, 0100,
_____| |_____
0010, 0001, 1000, 1100, 1110, ¦ 1111, 0111, . . .

In the second case, the output sequence has cycle length 15. The circuit of Figure 5.11(b) as said to generate a *maximal length pseudorandom binary sequence*. This is because the sequence of shift register contents includes each of the $2^n = 2^4$ possible 4-bit patterns except the 0000 pattern. As a result, the output is a sequence with period

$$p = 2^n - 1. \tag{5.7}$$

In the case of a 4-bit feedback shift register, the maximum sequence length is

$$2^4 - 1 = 15.$$

In general, the feedback structure for the n-stage shift-register must be chosen so as to produce a maximal length PRBS signal with sequence length (period) p given by Equation (5.7).

The circuit in Figure 5.11(a) does not satisfy this requirement. It illustrates how a particular choice of feedback structure can result in a partioning of the set of 2^n possible register states into cycle sets of nonmaximal length. For that example, there are four different cycle sets, two being of length 6, one of length 3, and one of length 1.

The cycle set sequences of operating states for Figure 5.11(a) are determined by the initial register contents. If the initial register states are 1111, we obtain the sequence:

$$1111, \; 0111, \; 0011, \; 1001, \; 1100, \; 1110, \; \ldots .$$

If the initial register contents were set to another state which is not in the above set, say 1010, this results in the following sequence being generated:

$$1010, \; 0101, \; 0010, \; 0001, \; 1000, \; 0100, \; \ldots$$

Likewise, if the initial state is 1011 say, we obtain the sequence

$$1011, \; 1101, \; 0110, \; \ldots ,$$

and finally, the only remaining cycle set is the single state 0000, $\ldots$.

For a *maximal length* sequence generator it is necessary to choose an appropriate feedback arrangement to ensure that only the two cycle sets are possible, one being the maximal length sequence and the other being the 0000 cycle set. We have seen that the feedback structure is specified by a polynomial of the form

$$x_1^1 = a_1 x^1 + a_2 x^2 + \ldots + x^n \tag{5.8}$$
$$\text{where } a_i = 0 \text{ or } 1 \text{ for } i = 1, 2, \ldots, (n - 1).$$

It can be shown that the required polynomial must be chosen to be one of a class of polynomials known as *irreducible characteristic polynomials* used in the theory of cyclic codes (discussed in Chapter 4, Volume 2). For further details on these polynomials, see for example Lin and Costello (4).

Each characteristic polynomial is of the form

$$g(X) = 1 + a_1 X + a_2 X^2 + \ldots + a_{n-1} X^{n-1} + X^n. \tag{5.9}$$

The values of a_1 to a_{n-1} are either 0 or 1 and indicate the feedback structure connections required for a given value of n.

Table 5.1 lists feedback polynomials which produce maximal length sequences from n-stage shift registers. Polynomials are given for values of n from 2 to 20. Also shown in the table are the sequence cycle lengths. It should be noted that, for each value of n, the polynomial given in the table represents only one choice of several possible polynomials which will produce a maximum length sequence. The polynomials listed in the table have the minimum number of terms, and hence, the minimum number of feedback connections. However, the table indicates the total number of polynomials for each value of n, which can be found to produce maximal length sequences.

Table 5.1
PRBS Sequence Generator Parameters

Number of shift reg. elements n	Characteristic (Feedback) polynomial $g(X)$	Sequence length $2^n - 1$	Number of different maximal length sequences
2	$1 + X + X^2$	3	1
3	$1 + X + X^3$	7	2
4	$1 + X + X^4$	15	2
5	$1 + X + X^5$	31	6
6	$1 + X + X^6$	63	6
7	$1 + X + X^7$	127	18
8	$1 + X^2 + X^3 + X^4 + X^8$	255	16
9	$1 + X^4 + X^9$	511	42
10	$1 + X^3 + X^{10}$	1023	60
11	$1 + X^2 + X^{11}$	2047	176
12	$1 + X + X^2 + X^{10} + X^{12}$	4095	144
13	$1 + X + X^2 + X^{12} + X^{13}$	8191	630
14	$1 + X + X^2 + X^{12} + X^{14}$	16383	1000
15	$1 + X + X^{15}$	32767	1800
16	$1 + X^2 + X^3 + X^5 + X^{16}$	65535	2100
17	$1 + X^3 + X^{17}$	131071	7600
18	$1 + X^7 + X^{18}$	262143	8000
19	$1 + X + X^2 + X^5 + X^{19}$	524287	19000
20	$1 + X^3 + X^{20}$	1048575	17000

Exercise 5.5

Design a feedback shift register structure which will produce a pseudorandom binary sequence of length 1023 bits.

Solution. From Table 5.1, we obtain for $n = 10$, the characteristic polynomial

$$g(X) = 1 + X^3 + X^{10}$$

from which we obtain the structure shown in Figure 5.12.

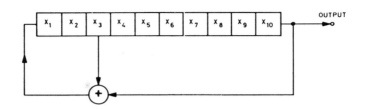

Figure 5.12   10-stage pseudorandom sequence generator.

5.4.4   Properties of Pseudorandom Binary Signals

We will now examine the characteristics of maximum length pseudo-random binary signals in some detail. A typical sequence waveform is shown in Figure 5.13. In order to describe and compare pseudorandom and random signals their statistical properties must be specified. For comparison purposes, the auto-correlation function is most convenient. In the following, we will compare the properties of PRBS waveforms with random noise and with random binary sequences. The discussion indicates that observations of long PRBS signals for times less than one sequence length suggest that the signal behaves as a truly random binary signal. It is only when observation times exceed one sequence length (T) that an indication of nonrandomness may be detected.

Autocorrelation function: Consider the *pseudorandom binary* signal $x(t)$ illustrated in Figure 5.13. For the waveform $x(t)$ the autocorrelation function is

$$R_{xx}(\tau) = \begin{cases} V^2 \left\{ 1 - |kT - \tau| \dfrac{p+1}{pT} \right\} & \text{for } |kT - \tau| < \Delta t \\[2mm] -V^2/p & \text{for } |kT - \tau| > \Delta t \\[2mm] & k = 0, +1, +2, \ldots \end{cases} \qquad (5.10)$$

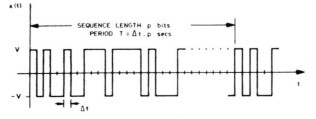

Figure 5.13   Typical PRBS waveform sequence.

where V, $-V$ are the PRBS signal amplitudes, p is the sequence length in bits, and $T = p.\Delta t$ is the period of the sequence. For details on the derivation of (5.10), see for example Golomb (8).

The result is illustrated in Figure 5.14(a). The autocorrelation function is periodic with period T. Within any one period length, it is identical in form to that of the *random binary signal* given by Equation (2.18) and illustrated in Figure 5.14(d). Because of this relationship, we can say that, observations of PRBS waveforms for times less than one sequence length, suggest that the signal is random.

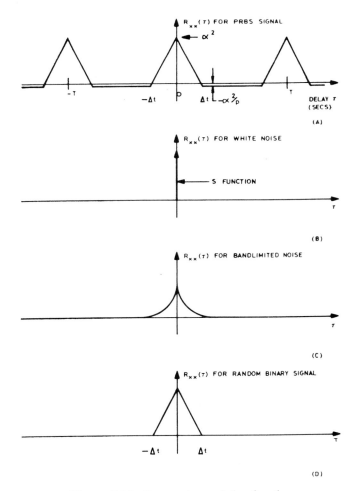

Figure 5.14 Some autocorrelation functions.

It is also of interest to compare the autocorrelation function of the PRBS signal with that of a *white noise* signal with zero mean. For the white noise signal, the autocorrelation function is a delta function at the time origin as illustrated in Figure 5.14(b). That is, for any value of relative delay other than zero, the white noise waveform and its delayed version are uncorrelated. If the noise is band-limited by a first-order low pass filter, its autocorrelation function will resemble that of Figure 5.14(c).

For PRBS signals with long sequence lengths p, the autocorrelation is a very small constant $(-V^2/p)$ for values of delay

$$\Delta t < \tau < (p - 1)\Delta t.$$

This is because the number of 1's in each period of the sequence is one greater than the number of 0's (because the all-zero register state is the one state that does not occur).

Frequency Spectrum: The frequency spectral characteristics can now be readily found. As discussed in Chapter 2, the power spectral density $S_x(f)$ of a random signal is the Fourier transform of its the autocorrelation function. Hence, for the PRBS waveform with autocorrelation function given by Equation (5.10), the spectral density is given by

$$S_X(f) = \frac{V^2}{p^2} \delta(f) + \frac{V^2(1+p)}{p^2} \sum_{\substack{i=-\infty \\ i \neq 0}}^{\infty} \mathrm{sinc}^2\left(\frac{i}{p}\right) \delta\left(f - \frac{i}{T}\right) \quad (5.11)$$

This is illustrated in Figure 5.15. The spectrum has an envelope of the form $\{(\sin x)/x\}^2$ where

$$x = \pi f/f_c.$$

Figure 5.15(a) illustrates the spectrum for the case where the PRBS sequence length is $p = 15$ bits. In Figure 5.15(b) the same spectrum is shown using logarithmic scales to show how the spectrum envelope varies in amplitude using a dB scale.

Figure 5.15(c) shows how the spectral lines become more closely spaced when the sequence length is increased. In this case, a 10-stage shift register is assumed so that the sequence length is $p = 1023$ bits. The clock frequency remains unchanged in each case.

The following important features of the spectrum of a PRBS waveform can be noted:

(1) the spectrum consists of discrete harmonically related components (because the original signal is periodic). By contrast, the power spectrum for a random signal is continuous.

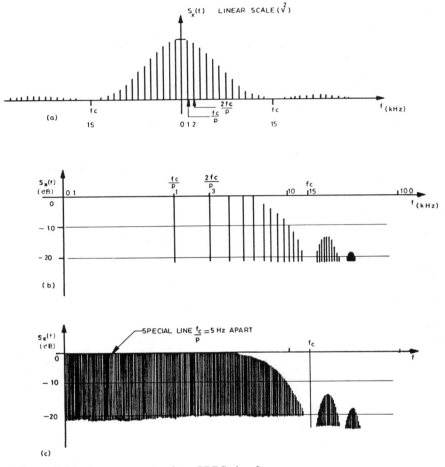

Figure 5.15 Power spectrum for a PRBS signal.
(a) Linear frequency scale; $p = 15, f = 15$ kHz (2 sided spectrum).
(b) Log f scale; $p = 15, f = 15$ kHz.
(c) Log f scale; $p = 1023, f = 15$ kHz.

(2) the spectral components occur at intervals of

$$1/T = f_c/p = 1/(p.\Delta t).$$

(3) the spectral envelope has nulls occurring at multiples of the clock frequency, namely at frequency values

$$nf_c = n/\Delta t.$$

(4) the amplitudes of the spectral components are substantially constant up to approximately $f_c/20$. This is of interest when the PRBS signal is to be used to approximate to a white noise test signal. If the PRBS waveform is passed through a low-pass filter with bandwidth approximately $f_c/20$, the resultant output signal will be a good approximation to a white noise signal at frequencies well below the filter cutoff frequency. To show how the spectral envelope of the PRBS waveform falls off in amplitude, the following values can be calculated from Equation (5.11)

Frequency	$S_x(f)$ (dB)
$0.05f_c$	-0.0358
$0.085f_c$	-0.10
$0.45f_c$	-3.00

(5) in comparison, the power spectrum of a *random* binary sequence will be a continuous one of the same form as the envelope above.

5.4.5   Applications of Pseudorandom Sequences

Pseudorandom binary sequences are used to approximate random sequences in measurements of eye-diagrams and bit error rates in digital transmission systems. PRBS waveforms have an autocorrelation function which is highly peaked for zero delay and approximately zero for other delays. Thus, they also find application where waveforms from remote sources must be synchronized. These applications include:

(1) word synchronization
(2) determination of range in a radar system
(3) measurement of the impulse response $h(t)$ of a linear system by cross correlation of input with output. This is called "system identification."

For details on these applications, see, for example, Ziemer (6).

The spectrum of a PRBS waveform closely resembles that of white noise for frequencies up to approximately 10 percent of the clock frequency, as illustrated in Figure 5.15(c). A suitable low-pass filter can, therefore, convert a PRBS waveform into one which closely resembles a low-pass Gaussian noise signal. This has applications in

(1) measurement of the transfer function of linear systems
(2) as a noise loading signal for intermodulation measurements over analogue radio systems.

PRBS sequences can also be used in the following areas:

(1) in encryption, as secure and limited access code generators
(2) in range radar applications in environments with high background noise, where a PRBS modulated pulse-train or CW signal has the property that its auto-correlation function is recoverable despite a noise-to-signal ratio in excess of many decibels
(3) in *spread spectrum* radio and radar systems designed to provide secure communication in a hostile environment
(4) in *simulation systems* as a means of generating random numbers.

Further details on these applications can be found in Diffie (7), Golomb (8), and Dixon (9).

5.5   ERROR RATE MEASUREMENTS

5.5.1   Bit-Error Rates

A common parameter used to describe the occurrence of errors in digital systems is the *long-term mean bit-error rate* (BER). In practice, this is measured by counting the number of information bits received in error N_E and comparing with the total number of information bits N_I received during a specified interval, to give

$$\text{BER} = N_E/N_I. \tag{5.12}$$

This value is also used as an approximation to the probability of error.

Often the information digits are structured either into characters (typically eight bits long) or blocks (up to several thousand bits) prior to transmission. In this case, the *character error rate* (CER) or *block error rate* (BLER) can be used. In many systems, the presence of one or more bit errors at unknown locations in a character or block may make the whole unit unusable. In this case, the CER or BLER give an indication of the percentage of usable received information units. Such an indication is not available from the BER, since bursts or clusters of errors affecting relatively few characters or blocks are not distinguished from randomly occurring errors which would affect far more characters or blocks.

The usual procedure for measuring the bit-error rate is by transmitting over the circuit a maximal length pseudorandom binary sequence generated by an *n*-stage shift register circuit. The longer the sequence length ($2^n - 1$), the more likely is the detection of malfunctions associated with specific data patterns. Typically, sequences of lengths 63, 511, and 2047 bits have been used for low-speed services such as voice-band data. For medium and high-speed digital transmission systems, sequences of the order of a million bits (that is, $n \geq 20$) are commonly employed.

The sequence length is chosen as a compromise between the desire to obtain as close an approximation to randomness as possible, and the cost of the hardware required to generate the sequence and to synchronize the error detector.

The error detection procedure consists of generating at the error detector, an identical sequence to the one transmitted, synchronizing both sequences in time and counting the number of places in which they differ. Synchronization of the sequence generated in the error detector with the sequence received is referred to as *frame synchronization*. One method of achieving frame synchronization at the receiver is to "search" the received information for a sequence exactly matching the received test sequence (we assume that bit synchronization is ideal).

A correlator circuit illustrated in Figure 5.16 is one method of carrying out the search. The correlator consists of a multiplier and integrate-and-dump circuit. If the two sequences are synchronized, the output of the integrator will be a maximum. If the patterns are out of synchronization, the integrator output will be approximately zero. In this case, the error detector PRBS generator sequence can be shifted in time (by interrupting its driving clock for one bit interval) and then the correlation process rechecked. This search procedure continues until lock is found.

The above is a relatively complex process and can be badly affected by transmission errors. A more popular approach is to use a feedback shift register circuit in the error detector, wired as the inverse of the PRBS generator in the transmitter. This is known as an *automatically synchronized error detector*. One possible implementation is illustrated in Figure 5.17. It shows a three-stage shift register example. (This is a similar configuration to that used in scrambler/descramblers which are employed to minimize the probability that long strings of 0's or 1's are transmitted. See Chapter 1, Volume 2.)

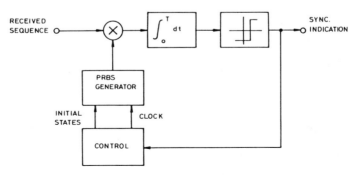

Figure 5.16 PRBS frame synchronization.

Exercise 5.6

Show by a specific example that, in the absence of transmission errors, the error detector circuit of Figure 5.17 will automatically establish synchronization within $n = 3$ cycles. For example, consider the initial state of the test generator shift registers to be 111 and that of the error detector to be 000. Show also what is the effect of a single channel error occurring to say, the eight transmission bit.

Solution. Table 5.2 shows the time sequence of operation of each of the shift register circuits at the sender and error detector, respectively.

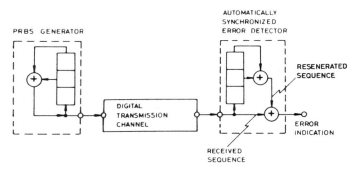

Figure 5.17   Error rate measurement using an automatically synchronized error detector.

Table 5.2
Error detector operation

Time	Shift reg. contents at sender	PRBS symbol sent	Channel Output	Shift reg. contents at error detector	Regenerated sequence	Error Indication
0	111	0	0	000	0	0
1	011	0	0	000	0	0
2	001	1	1	000	0	1
3	100	0	0	100(in sync)	0	0
4	010	1	1	010	1	0
5	101	1	1	101	1	0
6	110	1	1	110	1	0
7	111	0	0	111	0	0
8	011	0	1(error)	011	0	1
9	001	1	1	101	1	0
10	100	0	0	110	1	1
11	010	1	1	011	0	1
12	101	1	1	101	1	0

Note that for this method of synchronization and error detection, if an error occurs, this results in a temporary loss of synchronism. However, the detector circuit recovers rapidly from loss of synchronism.

However, when the erroneous bit enters the detector circuit, it causes error propagation. That is, each error in the received data will cause a number of false error indications to occur. For a single error isolated between correct bits, an error indication will be produced whenever the erroneous bit is tapped off from the detector synchronizer. Thus, if there are d taps in the synchronizer, each isolated error in the received sequence will introduce d error counts.

One way of avoiding erroneous counts with the automatic synchronizer of Figure 5.17 is to connect it to the error detector shown in Figure 5.18. This circuit is intended to remove the error propagation effect so that the decoded error pattern will be the original error pattern. The operation of Figure 5.18 can be illustrated by considering a specific example. This is done in the following exercise.

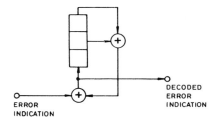

Figure 5.18   Decoder to avoid error propagation in automatic synchronizing error
detector.

Exercise 5.7

Investigate the performance of the decoder circuit in Figure 5.18 by tabulating the decoded error pattern for the error indication pattern generated in Exercise 5.6. Also investigate the effect of isolated double errors.

Solution. For the single error case, the decoded error pattern is as follows

Time	3	4	5	6	7	8	9	10	11	12
Error ind.	0	0	0	0	0	1	0	1	1	0
Decoded error	0	0	0	0	0	1	0	0	0	0

It is left to the reader to show that for the double error case, the decoded error pattern is also correct.

Many error rate test sets allow operation in either an automatically synchronizing mode or in a manual mode. Figure 5.19 shows an example of an error detector configuration that is manually synchronized. Synchronism is established by using the switch on position A to briefly connect the shift register input to the receiver output. Once the n shift register cells are loaded with error-free information bits (as indicated by an error detector output of zero), the switch is thrown to position B. Then the decoder synchronization will remain unaffected by subsequent errors in the receive sequence. As a result, only one error count will be produced by any one error.

However, if a bit is inserted or lost from the received sequence (for example, because of a slip), the decoder will not resynchronize and the operator must manually reload the decoder circuit using the switch.

In error rate test sets with manual or automatic operating modes, the manual mode is used for short tests, where accurate error rate estimates are needed and constant observation of the equipment is possible. For longer tests, where high accuracy is not important, the automatic mode is used. For further details see for example Newcombe (10).

To obtain statistically accurate estimates of the error rate, it is necessary to use a test sequence of sufficient length. Table 5.3 relates the number of errors counted and the minimum number of bits received during a test to the 99 percent confidence limits on the bit-error rate. The confidence limits indicate the range within which the error rate can be expected to lie with a probability of 0.99.

When the channel bit error rate is low, it may require a very long time to obtain an accurate estimate of the error rate. Typically, 20 error counts is a reasonable compromise for many purposes.

When error rates on a transmission system become very low, only one or two errors may occur in a day. Under these conditions, it becomes impractical to estimate the error rate by counting errors. When errors are extremely rare, it is usually adequate to state that the error rate is less than some specified value and only become concerned when the threshold is exceeded.

In these circumstances, while the absolute error rate may not be important, changes in the rate are of interest. A sudden increase may indicate that circuit

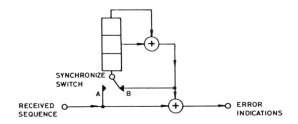

Figure 5.19 Manually synchronized error detector ($n = 3$).

Table 5.3
Variation of 99 percent confidence limits with number of errors recorded.

No. of errors in 10^6 received bits	Lower BER limit	Estimated BER	Upper BER limit
5	1×10^{-6}	5.0×10^{-6}	1×10^{-5}
10	5×10^{-6}	1.0×10^{-5}	2×10^{-5}
20	1×10^{-5}	2.0×10^{-5}	3×10^{-5}
40	3×10^{-5}	4.0×10^{-5}	5×10^{-5}
100	9×10^{-5}	1.0×10^{-4}	1.1×10^{-4}

failure is imminent. By sensing these conditions, a standby circuit can be switched in or some maintenance action scheduled.

A method of obtaining rapid estimates of low error rates is called *pseudo-error monitoring*. In this procedure, instead of monitoring the errors in the main transmission channel, one or more secondary receiver paths are deliberately added in which controlled amounts of degradation are introduced. As a result, the overall error rates of the data recovered via the secondary paths are higher than that in the main channel and thus can be measured faster. The error rates for several secondary paths are then extrapolated to estimate the rate in the main path. For details see Newcombe (10).

5.5.2 Error-Free Seconds

Theoretical studies of error performance in digital networks have been traditionally described in terms of a constant bit-error probability, numerically equal to the long-term mean BER. In part, the justification for this approach is based on the notion that a significant proportion of the errors in a real network arise as the result of purely random processes. In that case it could be assumed that the error occurrence statistics could be described in terms of a Poisson model (6).

That traditional BER approach has been found inadequate for describing network performance. This is because it does not provide an accurate characterization of the distribution of errors with time. In many practical digital systems, errors tend to occur in bursts or clusters to varying degrees. This is because many sources of noise or interference which cause errors are impulsive in nature. Also signal processing devices such as line coders and self-synchronizing scramblers can result in multiple errors for every single error on the transmission link.

The occurrence of a few very long error bursts can have a significant impact on the overall measured BER, but might not seriously affect a particular service in the digital network. Consequently, it is desirable to have alternative means of defining error performance in a way that more directly relates to the effect on services.

The approach being adopted in CCITT recommendations is to use two types of measure. They are

(1) the proportion of error-free time intervals thought to be useful for many data-type services
(2) the proportion of time that a specified short-term BER threshold is exceeded, based on circuit use for voice and some other services.

To express the performance of a data circuit in such a way that it is meaningful to both the network engineering organization and to the users, two parameters need to be used:

(1) *availability*—a data circuit is said to be available when its service is deemed to operate satisfactorily (that is, within prescribed error performance limits). Unavailability includes periods over which large numbers of data errors are encountered. These periods constitute what is known as outage time.
(2) *quality*—the performance of a circuit when it is available.

The *error-free second* (EFS) performance provides a useful basis for quantifying the performance of an available circuit. An error-free second is defined as a transmission time interval of one second in which no bit error is present. The EFS measure is, therefore, independent of the users' data rates and block sizes.

The circuit performance can then be expressed as a percentage of error-free seconds (%EFS). This percentage is obtained using the relationship

$$\%\text{EFS} = (\text{Number of EFS/Circuit available time}) \times 100. \quad (5.13)$$

The one second intervals which contain one or more errors are called *error seconds* (ES). Unavailability or outage time can be expressed in terms of error-seconds. For example, an event of 10 or more consecutive error-seconds may be defined as an *error-second outage* (ESO) or simply *outage*.

Test sets have been developed to permit evaluation of the availability and error performance of a data circuit in terms of the following parameters:

(1) Total number of error-free seconds (EFS) over a measurement period
(2) Total duration of error-second outages (ESO).

Since short-term error performance is as important as long-term error performance, the measurement period is subdivided into time intervals of 15 minutes.

Furthermore, in order to gain some insight into the transmission characteristics of the test circuit, the following information is recorded by the test set:

(1) The frequency distribution of the bit error counts for each error-second (BEC/ ES) within a 15-minute interval.

(2) The frequency distribution of error-free-second runs (EFSR) within the 15-minute interval above.
(3) The start and finish times (to the nearest millisecond) in real time of any externally detected alarm conditions.
(4) The start and finish times (to the nearest second) in real time of any events of 10 or more *consecutive* error-seconds (or error-second outages).

It is possible to express the bit error rate in terms of the percentage of error free seconds. The relationship is

$$\text{BER} = \frac{e}{r_b} \left\{ 1 - \frac{(\%\text{EFS}}{100} \cdot \frac{\%\text{Availability})}{100} \right\} \tag{5.14}$$

where

e = (total number of bit errors)/(Number of ES)

r_b = data bit rate

$$\% \text{ EFS} = \frac{\text{Error-free-second time}}{\text{Available time}} \times 100$$

$$\% \text{ Availability} = \frac{\text{Available time}}{\text{Valid time}} \times 100.$$

The valid time refers to the test period excluding those intervals where the circuit is disconnected for planned operational purposes. For details, see for example Duc (12). As a specific case, assume that the performance objectives of a data circuit are:

$$\% \text{ Availability} = 99.9\%$$

$$\% \text{ EFS} = 99.5\%$$

The equivalent BER objective as a function of e (average bit-error count per error-second), is shown in Figure 5.20 for bit rates of 64 kbit/s, 2048 kbit/s and 140 Mbit/s, respectively. Note that bit-error rate performance objectives at various data rates are achieved in the region on or below the respective limits.

For a given test data circuit operating at one of the above bit rates, the BER performance can be evaluated from measured parameters, namely, % Availability, % EFS, and e. This BER result can be then compared with the corresponding performance objective in Figure 5.20.

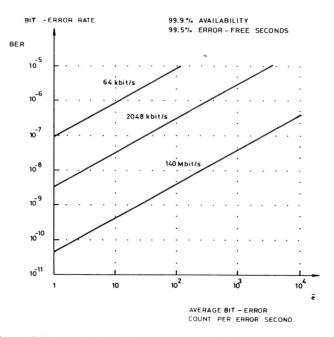

Figure 5.20 Equivalent bit-error rates at 64, 2048 and 140,000 kbit/s for

Availability = 99.9 % and
% Error-free seconds = 99.5%.

5.6 REGENERATOR FAULT LOCATION TESTS

5.6.1 Triples Test Signal

In a digital transmission line consisting of several regenerators in tandem, it is convenient to have a method of locating any faulty regenerators remotely from the terminal. The "trio" or "triples" fault location system is commonly used to achieve this.

In this fault location process, the traffic-generated HDB3 digital line signal must be taken off the system and replaced by a special supervisory test signal. This test signal is illustrated in Figure 5.21(a) and (b). It consists of a sequence of triples, each triple consisting of the successive symbols $+1$, -1, $+1$ (known as a positive triple) or -1, $+1$, -1 (known as a negative triple). The triples are spaced NT_b seconds apart, where N is an integer. The value of N can be varied at the test signal generator typically over the range 4–11. Such a pulse sequence is said to be a triples test signal with density of $1/N$. In Figure 5.21(a)

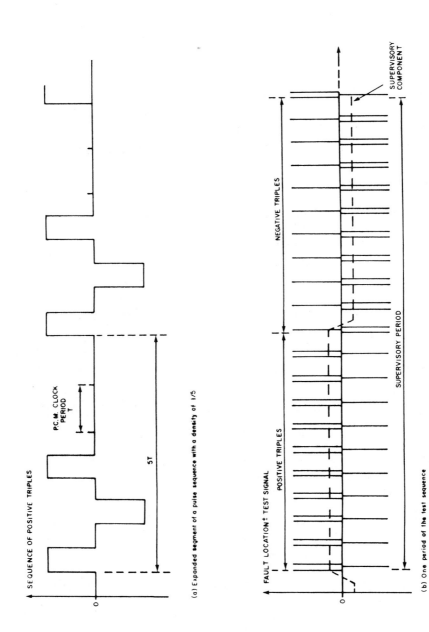

Figure 5.21 Triples test signal.

a sequence of positive triples is shown with $N = 5$. Each symbol is represented by a half-width rectangular pulse.

As illustrated in the compressed example in Figure 5.21(b), the test signal consists of positive triples alternating with sequences of negative triples. The length of each period T can be set at the signal generator to be one of 24 possible values. On examination of Figure 5.21(b), it is apparent that the triples test signal contains a voice frequency component, shown dotted in the diagram. The fundamental frequency of this component $1/T$ can, therefore, be set to one of 24 possible frequency values $f_1, f_2, \ldots, f_{24}$, typically in the range 1–3 kHz.

At each regenerator, an auxiliary output from the transmit amplifier is connected to a separate supervisory channel via a bandpass filter. This filter is known as a supervisory filter and is centered at one of the frequencies $f_1 - f_{24}$. Each regenerator location in a line system has a unique supervisory filter center frequency. The filter output signal can be returned to the terminal via the feedback supervisory channel, typically another cable pair.

5.6.2 Fault Location Procedure

The triples test signal enables an operator at the terminal to test each regenerative repeater in a particular line system. The regenerator under test is selected by setting the period of the test signal so that the supervisory signal fundamental component matches the particular center frequency of the regenerator's supervisory filter. If that regenerator and all those preceding it are operational, then the digital test signal will be regenerated. The audio component of this signal will be returned to the terminal via the filter at the regenerator under test. The magnitude of this return signal can be used to indicate whether or not the regenerator is fully operational. By progressively selecting test signal supervisory filter frequencies working outwards from the testing station, it is possible to locate a faulty regenerative repeater even in a long string of up to 24 of these units.

By increasing the density $1/N$ of the triples test signal, it is possible to increase the amplitude of this supervisory component in the test signal (shown dotted in Figure 5.21(b)). That is, the average value of the line signal consisting of a sequence of positive triples is decreased. If the regenerator is functioning correctly, the amplitude of the feedback supervisory signal should also increase. If the supervisory signal amplitude falls off, then the onset of a significant number of errors is indicated. In practice, it has been found that by varying the triples density, an indication can be obtained of any deterioration in the regenerator noise margin or errors in decision threshold level settings.

5.7   ERROR PERFORMANCE OBJECTIVES

International standards for performance of telecommunications networks are established by the CCITT (International Telegraph and Telephone Consultative Committee). The CCITT forms part of the International Telecommunication Union (ITU) based in Geneva, Switzerland. There are over 160 countries who are ITU Member countries.

The CCITT standards are published every four years (15). Each publication consists of a "Book" of approximately 30 volumes (termed fascicles) and is the outcome of a four-year study period. During each four-year period, approximately 15 Study Groups of international network operator and government representatives discuss, negotiate, and elaborate international standards in all areas of telecommunications (except for radio communications).

At the end of each four-year cycle the CCITT Plenary Assembly meets to give formal approval to these standards known as "Recommendations." Each Plenary Assembly also attaches a color name to the Book to distinguish it from previous versions. For example, the *Red Book* emanated from the decisions of the October 1984 Plenary Assembly (15).

5.7.1   Hypothetical Reference Connection (HRX)

Many digital network performance standards recommended by the CCITT are based on transmission system models known as a Hypothetical Reference Connections (HRX). For example, the longest length HRX model used by the CCITT describes an all-digital 64 kbit/s connection over multiple lengths of transmission and switching stages as illustrated in Figure 5.22. Recall that one 64 kbit/s connection represents one PCM telephone channel. The model represents an international circuit with the longest envisaged length, namely a 27500 km connection incorporating local connections at either end and national and international connections in between. Various types of switching centers are shown incorporated in the circuit. For details on these see for example CCITT (15). Digital switching principles will be discussed in Chapter 2, Volume 2.

The HRX model does not attempt to describe any particular real circuit. The diversity of national and international networks in respect of country size, circuit implementation, national policies, and so forth, naturally preclude the incorporation of all of the significant features into a single model, or even a few simple network models.

The longest length HRX provides a uniform basis for the study of impairments in long digital circuits. It provides a common reference datum for the specification of international performance standards. National and local administrations must then develop their own representative networks models reflecting the features

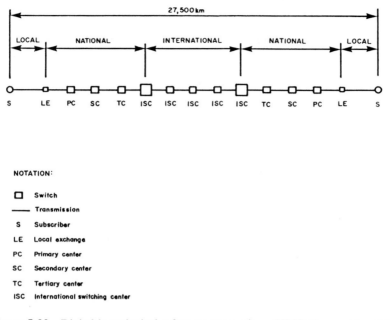

Figure 5.22   Digital hypothetical reference connections (HRX) (Longest length).

of their own evolving national digital networks. They can then use these to validate prima-facie compliance with international standards.

5.7.2   Error Performance Specifications

The original objective proposed for the longest length international 64 kbit/s HRX was in terms of a limiting value for the long-term mean BER. An overall value of 1×10^{-5} was taken to represent a level of impairment which was considered "just discernible on low level speech" when using PCM. However this objective was never adopted by the CCITT.

The performance objectives of the CCITT are aimed at network users. Because of this, the performance is specified in a way that gives the most information about the effects on the services provided by the network.

Although the objectives are intended to suit the needs of many different services, during the formulation of the CCITT recommendations, two important classes of users were identified, namely telephony users and data users. In formulating the objectives, it was necessary to reach a compromise between the desire to meet service needs, and the feasibility of realizing appropriate transmission systems taking into account economic and technical constraints.

The network performance objective for the longest international 64 kbit/s HRX from customer to customer is given in Table 5.4. It is expressed in three parts, all parts to be met concurrently. In regard to part (b) of Table 5.4, note that when the BER is worse than 1×10^{-3} for periods equal to or exceeding 10 consecutive seconds, the connection is considered to be *unavailable*. Part (c) of the recommendations is the error-free-seconds requirement.

It is necessary to subdivide the overall end-to-end objectives into constituent parts of networks and even to individual equipment and subsystems. This is so that the end-to-end objectives can form the basis upon which performance standards are derived for individual parts of an overall system.

The apportionment philosophy is based on the assumed use of transmission systems having qualities that fall into one of the following three grade classifications:

(1) *Local Grade*: This includes systems assumed to be operating between customers' premises and local exchanges, and typically low bit-rate systems operating at a rate below the primary multiplex levels of 1544 and 2048 kbit/s over a diverse range of cable types and other media.

(2) *Medium Grade*: This includes systems assumed to be operating between local exchanges and beyond, into the national part of the connection. The actual distance covered by such systems is expected to vary considerably from one country to another, but under most circumstances will not exceed

Table 5.4
Error performance objectives for an international HRX at 64 kbit/s.

Part	Objective
(a)	At least 90 percent of 10 minute intervals to have 38 or fewer errors (Note 1)
(b)	At least 99.8 percent of one-second intervals to have less than 64 errors (Note 2)
(c)	At least 92 percent of one-second intervals to have zero errors (Note 3)

Notes:
1. 38 errors over 10 minutes is equivalent to a BER of approximately 1×10^{-6}.
2. 64 errors over one second is equivalent to a BER of 1×10^{-3}.
3. The one-second integration period was chosen in the belief that it represented a short enough time interval to display sufficient detail of the error structure, yet long enough to encompass most block lengths that data users may choose for error control.

1250 km. Typically, these systems will operate at low, medium, or high bit rates and use a variety of media.

(3) *High Grade*: This includes long-haul transmission systems realizing national and international parts of a connection. Typically such systems are expected to operate at medium or high bit rates utilizing all types of media (for example metallic, optical, radio, and satellite.) Of the total 27500 km HRX connection length, a significant proportion will be implemented by using high grade plant.

Typically the allocation of the performance quality allowance over a 27500 km HRX is broken up into allowances for the following portions:

Local grade, medium grade	: the first 1250 km from a customer premises
High grade	: 2500 km national/international segments
Medium grade, local grade	: the final 1250 km to a customer premises.

The overall apportionment philosophy involves the use of two slightly different strategies. One is applicable to the 10-minute interval objectives (part (a) of Table 5.4) and to the error-free-second objectives (part (c)). The other is applicable to the availability objectives (part (b)).

Table 5.5 shows the allocation of 10-minute intervals and errored-seconds objectives. It is based on a 15 percent block allowance for the Local grade and Medium grade segments, respectively, at each end and 40 percent for the High grade central part. The latter is equivalent to a conceptual quality of 0.0016 percent per km for 2500 km.

Table 5.5
CCITT Allocation of percentage of 10-minute and errored-seconds objectives

| Quality classification | Network Performance Objective at 64 kbit/s | |
	% of 10-min intervals with $>$ 38 errors	% of seconds with one or more errors
Local grade—allowance for each end	$<$ 1.5	$<$ 1.2
Medium grade— allowance for each end	$<$ 1.5	$<$ 1.2
High grade—2500 km	$<$ 4.0	$<$ 3.2

The above apportionment strategy is considered to be inappropriate for the circuit availability (severely-errored-seconds) objective. The total allowance of 0.2 percent from Table 5.4 is subdivided between each classification (local, medium, and high) in the following manner:

(1) 0.1% is divided between the three circuit classifications in the same proportions as adopted for the other two objectives. This results in the allocation given in Table 5.6. The allocations of Table 5.6 are independent of the transmission media being used. In practical systems, the performance is dependent on the type of transmission medium. For example, terrestrial radio relay and satellite systems are subject to adverse propagation conditions. It is normally not economic to design such systems to provide a good error-ratio performance for 100 percent of the time but accept that occasionally as a result of extreme propagation conditions, the performance will be degraded and, by design, will impinge on the circuit availability allowance.

(2) In recognition of the difficulty experienced in controlling the occurrence of severely-errored seconds in propagation limited systems, particularly during worst month conditions, the remaining 1 percent of the allowance from Table 5.4 is allocated to those types of systems falling into the medium and high-grade classifications. The allocation takes into account the fact that adverse propagation conditions are unlikely to be experienced simultaneously over all parts of the HRX. The allocated percentages are

2500 km radio system : an additional 0.05%

medium grade section : an additional 0.05%

satellite link   : an additional 0.01%.

Table 5.6
CCITT Allocation of availability objectives

Quality classification	Network performance objectives at 64 kbit/s: % of seconds with 64 or more errors.
Local grade (each end)	< 0.15%
Medium grade (each end)	< 0.15%
High grade (25000 km)	< 0.04%

An interesting survey and further details of the development of CCITT error performance objectives is given in McLintock (14). The CCITT objectives are given in (15). Additional information on error measurement and associated instrumentation is given in Huckett (16) and Duc (11).

5.8 QUESTIONS AND PROBLEMS

5.1 Sketch the eye pattern for a noise-free line waveform composed of a AMI line code with basic pulse shape

$$p_r(t) = \begin{cases} 2t/T_b & , & 0 < t < 0.5T_b \\ 1.25 - 0.5t/T_b, & 0.5\ T_b < t < 2.5\ T_b \\ 0 & , & \text{otherwise.} \end{cases}$$

5.2 Repeat Question 5.1 for an HDB3 line code.

5.3 Sketch the eye pattern for a noise-free line waveform composed of an AMI line code with basic pulse shape given by Equation (4.19).

5.4 Repeat Question 5.3 for the case where the basic pulse shape is given by the Nyquist shaped pulse of Equation (3.13) with $\alpha = 1.0$.

5.5 Sketch the eye pattern for a noise-free duobinary signal with basic pulse shape given by Equation (3.30)

5.6 Determine values for the noise margin degradation resulting from ISI for each of the cases described in Questions 5.1 to 5.5.

5.7 Draw a block diagram of the system required to display the eye pattern at the regenerator sampler input of a line transmission system designed to carry 2 Mbit/s PCM signals.

5.8 Find the pseudorandom binary sequence (PRBS) output from a five-bit feedback shift register with characteristic feedback polynomial given by

$$g(X) = 1 + X^3 + X^5.$$

Assume the shift register contents are initially set to all 1's.

5.9 Repeat Question 5.8 for the case where

$$g(X) = 1 + X + X^5.$$

5.10 Design a feedback shift register structure which will produce a pseudo-random binary sequence of cycle length of at least 10^6 bits.

5.11 Consider a method of generating a pseudorandom binary sequence $\{a_k\}$ using a calculator or personal computer.

(1) Use the recursive formula

$$x_k = \text{fractional part of } \{(\pi + x_{k-1})^5\}$$

to generate pseudorandom members in the range $0 \leq x < 1$.

(2) Obtain a binary sequence $\{a_k\}$ from the x_k values.

(3) Carry out some computations to test the randomness of the sequence $\{a_k\}$.

5.12 Use a random number generator subroutine on a main frame or personal computer to generate a pseudorandom binary sequence $\{a_k\}$ containing 100 values. Carry out some computations to test the randomness of the sequence.

5.13 Consider the pseudorandom binary sequence generated by the 4-bit shift register circuit shown in Figure 5.10(b). If the clock frequency is 2 Mbit/s, sketch and dimension the power spectrum of the sequence in dB versus frequency.

5.14 The 10-stage pseudorandom binary sequence generator shown in Figure 5.11 is to be used to test the error rate on a digital transmission system with binary input and output.

(1) Design an automatically synchronized error detection circuit suitable for use at the receiving end of the system. Include a suitable decoder to avoid error propagation.

(2) Show by an illustrative example that an isolated error will not cause erroneous error counts because of error propagation.

5.9 REFERENCES

1. G.J. Semple, "The Effect of Intersymbol Interference on the Operation of PCM Line Regenerators," *ATR Australian Telecommunications Res.*, Vol. 12, No. 1, pp. 17–31, 1978.

2. K. Feher, "Digital Modulation Techniques in an Interference Environment," *EMC Encyclopedia*, Vol. 9, Ch. 4, Don White Consultants, Gainesville, VA, 1977.

3. A.J. Gibbs, "Measurement of PCM Regenerator Crosstalk Performance," *I.E.E. Electronics Letters*, Vol. 15, No. 3, pp. 82–83, 1979.

4. S. Lin and D.J. Costello Jr, *Error Control Techniques: Principles and Applications*, Prentice-Hall, 1983.

5. M.G. Hartley, *Digital Simulation Methods*, Peter Peregrinus Ltd., 1975.

6. R.E. Ziemer and W.H. Tranter, *Principles of Communications Systems, Modulation and Noise*, Houghton Mifflin, 1976.

7. W. Diffie and M.E. Hellman, "New Directions in Cryptography," *I.E.E.E. Trans. Inf. Theory*, Vol. IT-22, No. 6, pp. 644–654, 1976.

8. S.W. Golomb (Ed.), *Digital Communications—With Space Applications*, Prentice-Hall, 1964.

9. R.C. Dixon, *Spread Spectrum Systems*, Wiley, 1976.

10. E.A. Newcombe and S. Pasupathy, "Error Rate Monitoring for Digital Communications," *Proc. of I.E.E.E.*, Vol. 70, No. 8, pp. 805–828, August 1982.

11. N.Q. Duc, R.B. Coxhill, K.S. English, and R.I. Webster, "A Performance Monitoring System for Data Transmission Circuits," *Telecommunications Journal of Australia*, Vol. 30, No. 2, pp. 110–116, 1980.

12. N.Q. Duc, "Estimation of Bit Error Rate from Error-free Second Performance," *ATR Australian Telecommunications Res.*, Vol. 13, No. 2, pp. 50–52, 1980.

13. G.J. Semple and A.J. Gibbs, "Assessment of Methods for Evaluating the Immunity of PCM Regenerators to Near End Crosstalk," *IEEE Trans. Comm.*, Vol. COM-30, No. 7, pp. 1791–1797, July 1982.

14. R.W. McLintock and B.N. Kearnsey, "Error performance objectives for digital networks," *The Radio and Electronic Engineer*, Vol. 54, No. 2, pp. 79–85, Feb. 1984.

15. CCITT (International Telegraph and Telephone Consultative Committee), VIIth Plenary Assembly Yellow Book (1980) and VIIIth Plenary Assembly Red Book (1984), International Telecom. Union, Geneva.

16. P. Huckett, "Performance evaluation in an ISDN—digital transmission impairments," *The Radio and Electronic Engineer*, Vol. 54, No. 2, pp. 96–106, Feb. 1984.

Chapter 6

DIGITAL CODING OF SPEECH SIGNALS

This chapter examines the development of efficient methods for conversion of analogue signals into binary sequences. A variety of procedures exist for digital encoding of speech, music, television, and other analogue signals. The essential features of speech coding schemes will be discussed and their performance compared.

6.1 INTRODUCTION

6.1.1 Coding Methods

Digital speech encoding techniques can be broadly classified into two types.

(1) *Waveform coding* involves the analogue-to-digital conversion of analogue waveforms as faithfully as possible. Typical methods are pulse code modulation (PCM), adaptive differential pulse code modulation (ADPCM), delta modulation (DM), and linear predictive coding (LPC).

(2) *Source model coding* involves the modelling of the vocal system to generate a digital signal representing the perceptually significant aspects of the speech production and hearing processes such as voiced/unvoiced segments, pitch, and vocal filtering. A device using this class of techniques is generally referred to as a vocoder. (Voiced segments of speech are those regions in which the vocal cords are vibrating and generating a pulse-like sound. Unvoiced segments are noise-like intervals caused by air turbulence at points of constriction in the vocal tract.)

Waveform coding techniques of varying degrees of complexity can be employed to obtain good quality speed transmission at bit rates from 64 kbit/s (for example, in conventional PCM) down to about 10 kbit/s (using APC techniques).

By using source model coding (a vocoder), the bit rate for intelligible speech can be reduced to approximately 2 kbit/s. Vocoder techniques tend to produce a synthetic, machine-like unnatural quality so they are not generally considered suitable for public telephone transmission. In this chapter, we will concentrate on waveform coding techniques. For further information on source model coding, see for example, Flanagan (3).

Pulse code modulation (PCM) represents the most common method of encoding speech for digital transmission in telephone networks. The essential features of PCM systems have been outlined in previous chapters. In this chapter, the factors that limit the transmission quality of PCM systems are examined in the context of international standards for public networks.

In conventional PCM systems, we have seen that each encoded telephone speech signal generates a 64 kbit/s digital bit stream. Many other encoding techniques have been developed in an attempt to achieve acceptable quality transmission at lower bit rates. These techniques include delta modulation (DM), differential pulse code modulation (DPCM), and linear predictive coding (LPC), which take advantage of known spectral characteristics (local time correlation) of speech waveforms.

Figure 6.1 illustrates the general procedure for waveform coding of digital signals and reconstruction of the analogue output after transmission. The key operations can be summarized as follows.

Sampling transforms a continuous time signal into a discrete set of samples, uniformly spaced in time. The values of the samples can lie in a continuous amplitude range. The sampler output is referred to as a *discrete-time* (or sampled-data) signal.

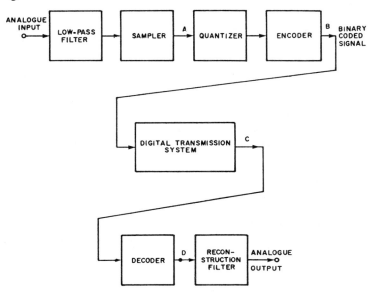

Figure 6.1   Basic elements of encoding and decoding systems.

Quantization is the process of representing a continuous amplitude variable by a finite number of distinct values. In speech encoding, the quantization process is necessary in order to establish a finite set of amplitude levels, each of which can then be encoded into fixed-length binary code word. The quantizer output is known as a *digital* signal, that is, one in which both time and amplitude are discrete-valued. The quantization process introduces some error into the encoded signal values. The resulting distortion is referred to as *quantization noise*. As we shall see, quantization noise imposes an important constraint in the trade-off between encoder bit rates and the quality of the speech delivered to the listener.

Encoding is the process of converting the quantized signal amplitudes into binary symbols. After transmission, *decoding* converts the binary signal into a set of discrete-time samples again. Then *reconstruction* is the procedure by which the continuous signal is recovered from the samples.

The processes of sampling, quantization, and binary encoding can be thought of as *source coding*. This was referred to in Chapter 2. In general, efficient source coding procedures attempt to take advantage of some knowledge of the nature of the input analogue waveform to generate binary signals with minimum bit rates for given distortion critieria. In some waveform coding techniques, such as DPCM, bit rates are reduced by taking advantage of the high correlation between adjacent sample values.

6.1.2   Distortion Criteria

In most schemes for the digital transmission of analogue signals, there are three sources of distortion giving rise to the following types of distortion errors in the analogue signal recovered after transmission:

(1) *Sampling and reconstruction errors*—even if the sample values are encoded, transmitted, and reproduced exactly at the receiving end, the output analogue waveform after reconstruction filtering may not be the same as the input analogue waveform. The amount of distortion depends on the sampling rate and reconstruction strategies.

(2) *Quantization distortion*—the process of representing analogue sample values by a binary code signal involves the conversion of continuous amplitude samples to a finite number of discrete amplitude values. Quantization distortion is the term given to the error that results from the rounding offs involved in attempting to represent a signal with a finite set of numbers. The amount of distortion depends on the number of quantization levels and the amplitude probability distribution of the samples of the analogue signal.

(3) *Channel errors*—during transmission, errors may result from channel noise, crosstalk, or intersymbol interference. Errors that remain can result in received signal quantization levels that differ from the corresponding trans-

mitted levels. These give rise to a further source of waveform distortion after reconstruction. The amount of distortion depends on the channel bit error rate, which was discussed in Chapters 2–4.

In order to quantify the effects of these different types of distortion or errors on the received waveform, we need a criteria of "goodness." We need to decide what is to be the basis for assessing whether or not one coding scheme is better than another? A reasonable answer to this question is that we should seek "the scheme with the least error." However, this only leads to another question: "What is a reasonable definition for error?"

In an attempt to arrive at a reasonable error criterion, we might start with the waveform error, that is, the error between the input analogue waveform $x(t)$ and output waveform $\hat{x}(t)$. This is,

$$e(t) = x(t) - \hat{x}(t). \tag{6.1}$$

Suppose now that we put the same waveform $x(t)$ into two different telemetry systems, say system A and system B. Then we might observe the waveform errors $e_1(t)$ and $e_2(t)$ as illustrated in Figure 6.2. Which system has the smaller error?

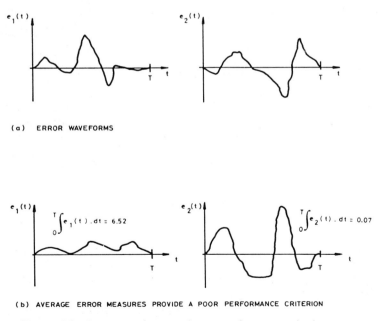

(a)   ERROR WAVEFORMS

(b) AVERAGE ERROR MEASURES PROVIDE A POOR PERFORMANCE CRITERION

Figure 6.2   Error waveforms and error performance criteria.

It is obvious that the waveform error at every instant of time is too detailed a measure of error. In search of a somewhat more gross error measure, we might try the integral of the error

$$\int_0^T e(t)dt = \int_0^T [x(t) - \hat{x}(t)]dt. \tag{6.2}$$

However, this is not a satisfactory performance measure because it allows positive errors to cancel the effect of negative errors, as illustrated in Figure 6.2(b).

We can get around this problem by considering the integral of the magnitude of the error or the integral of some even power of the error. For example, we might try the integral magnitude (im) error or the integral square (is) error, that is

$$\text{im error} = \int_0^T |x(t) - \hat{x}(t)| dt \tag{6.3}$$

$$\text{is error} = \int_0^T [x(t) - \hat{x}(t)]^2 dt. \tag{6.4}$$

Both of these error criteria have been used for analysis of digital transmission systems, but the second is far more tractable in most cases. That is, it is mathematically more convenient in that it is much easier to compute. In practice, it is usual to treat the error waveform as though it were noise, and to compute the *mean-square* noise voltage (noise power in 1 ohm) using

$$N = \lim_{T \to \infty} \frac{1}{T} \int_0^T [x(t) - \hat{x}(t)]^2 dt. \tag{6.5}$$

Finally, note that the mean square error depends on the particular input $x(t)$. Hence, for one input waveform, system A may have a smaller mean square error than system B, whereas, for a different input waveform system B may have the smaller error. Hence, it is sometimes necessary to consider the average of the integral square error over all input waveforms. That is called the mean integral square (mis) error

$$\text{mis error} = E\left\{\frac{1}{T} \int_0^T [x(t) - \hat{x}(t)]^2 \, dt\right\} \tag{6.6}$$

where the expectation operator $E\{.\}$ is with respect to the random process $x(t)$.

Mathematically speaking, the mean square error is a very convenient performance criterion for comparing PCM and other coding systems, because it is easier to compute than most other error criteria. On the other hand, the relationship between this objective criterion and any subjective criteria is not at all clear. For example, in television pictures for entertainment, there does not seem to be much correlation between the number given by the mean square error criterion and how pleasing the picture is "to look at" as determined by subjective evaluations by a number of observers. The most plausible explanation is that large errors are severely penalized by the mean-square error (because of the square), while small errors are hardly penalized at all. For example, we can visualize two pictures having the same mean square error. The first picture has a small amount of "snow" uniformly distributed over the whole picture. The second picture is perfectly clear except for a few large specks. When asked which is the "better picture," most observers would choose in favor of the latter.

On the other hand, in the case of facsimile, the mean square error appears to be a reasonable criterion. For example, when transmitting a document such as a financial statement, we are more interested in the information conveyed than in how pleasant the received document is to look at. A uniformly hazy, but readable document is preferred to one that is very clear except that one or two numbers are smudged to the extent that they cannot be recognized.

The perfect error criterion for the above applications would not only take into account the purpose of the picture, but also the physiological characteristics of the human eye. Certain types of errors are automatically filtered out by the observer. These need to be penalized less heavily than other errors that are more bothersome to the average observer.

Similar arguments hold for speech waveforms. For example, the performance criterion suitable for a communication system to convey a message from an aircraft pilot to a ground based air traffic controller is quite different from that for a system transmitting soft, soothing music to relax by. Again, some types of errors are more bothersome than others because of the characteristics of the human ear.

We will use the mean square error criterion because it is mathematically convenient. It is also sometimes a reasonable performance measure. Most important, we know of no better criterion. We must, however, keep in mind its shortcomings and not blindly accept all of its implications. Ultimately, it is necessary to build a system and subject it to detailed subjective performance evaluation.

In the next sections, we will examine causes of distortion in conventional PCM systems used for telephone speech.

6.2 IMPAIRMENTS RESULTING FROM SAMPLING

In this section, we consider the possible impairments that may result from the sampling and reconstruction procedures. To isolate these impairments from others, we initially assume that the encoding, transmission, and decoding procedures operate in an ideal manner. That is, in Figure 6.1, the amplitude value of each

sample at the encoder input (point A) is recovered exactly at the decoder output
(point D).

6.2.1 Flat-top Samples

Figure 6.3(a) illustrates a sampling process in which a continuous analogue signal
$x(t)$ is transformed into a sequence of samples spaced $T = 1/f_s$ sec apart, where
f_s is the sampling rate. The sampled signal is denoted $x_s(t)$. These samples are
assumed to be encoded, transmitted, and decoded without any change to the
amplitude values. The input to the reconstruction circuit is denoted $y_s(t)$. This

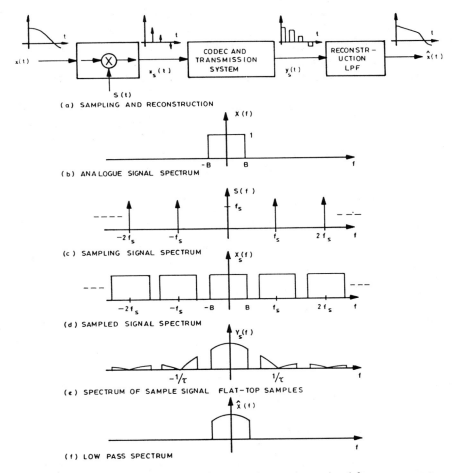

Figure 6.3 Sampling and reconstruction operations and associated frequency spectra.

is assumed to be a sequence of the same sample values as in $x_s(t)$. In practice, it is likely that each sample is represented by a pulse of width τ where

$$\tau \leq T_s, \qquad T_s = 1/f_s.$$

It is useful to examine the frequency spectrum at each point in the system, as illustrated in Figure 6.3(a). Consider the analogue signal $x(t)$ to be a baseband signal with frequency components in the range $0 - B$ (Hz). Its Fourier transform $X(f)$ is illustrated in Figure 6.3(b).

The sampled signal $x_s(t)$ can be represented algebraically as

$$x_s(t) = x(t).s(t) \tag{6.7}$$

where $s(t)$ is a sequence of delta functions. The sequence $s(t)$ is called a sampling function and is defined

$$s(t) = \sum_{k=-\infty}^{\infty} \delta(t - kT_s). \tag{6.8}$$

The sampling operation can be conveniently envisaged as the multiplication of the original analogue signal $x(t)$ by either 1 (at the instant when it is being sampled) or 0 (at other times when the sampler output $x_s(t)$ is zero). Hence, we can write

$$x_s(t) = x(t) \sum_{k=-\infty}^{\infty} \delta(t - kT_s)$$

$$= \sum_{k=-\infty}^{\infty} x(kT_s).\delta(t - kT_s).$$

The periodic sampling function $s(t)$ has the Fourier Transform

$$S(f) = f_s \sum_{k=-\infty}^{\infty} \delta(f - kf_s) \tag{6.9}$$

which represents an infinite number of harmonically-related frequency components each of amplitude f_s as illustrated in Figure 6.3(c).

To find the spectrum $X_s(f)$ of the sampled signal $x_s(t)$, we note the form of Equation (6.8) and recall that *multiplication in the time domain is equivalent to convolution in the frequency domain*. That is, we can write the spectrum of the sampled signal as

$$X_s(f) = X(f) * S(f) \tag{6.10}$$

where $*$ denotes the convolution operation.

The convolution of $X(f)$ with any one of the delta functions,

$$\delta(f - kf_s)$$

results in a shift of the original spectrum by an amount kf_s, resulting in the spectrum $X(f - kf_s)$. Using superposition, we can sum the results of the convolution to obtain

$$X_s(f) = \sum_{k=-\infty}^{\infty} f_s X(f - kf_s). \qquad (6.11)$$

As shown in Figure 6.3(d), the spectrum of the sampled signal is an infinite sequence of images of the original spectrum, spaced f_s apart and each weighted by the constant f_s.

To reconstruct the analogue waveform, it is obvious from Figure 6.3(d) that all that is required is a low-pass filter with bandwidth B to recover the original components $X(f)$. However, $X_s(f)$ is not the spectrum of the waveform assumed present at the reconstruction low-pass filter input in Figure 6.3(a). At this point, the waveform $y_s(t)$ is a sequence of flat-top samples

$$y_s(t) = \sum_{k=-\infty}^{\infty} x(kT_s).p(t - kT_s) \qquad (6.12)$$

each sample being of amplitude $x(kT_s)$ and having the basic rectangular pulse shape

$$p(t) = \begin{cases} 1, & 0 < t \leq \tau \\ 0, & \text{otherwise.} \end{cases} \qquad (6.13)$$

Now we can observe that the equation for $y_s(t)$ has the form of a convolution of the basic rectangular pulse $p(t)$ with the instantaneous samples of $x(t)$. That is, we can write

$$y(t) = x_s(t) * p(t)$$

$$= \left[\sum_{k=-\infty}^{\infty} x(kT_s).\delta(t - kT_s) \right] * p(t).$$

This convolution in the time domain is equivalent to multiplication of the transforms in the frequency domain. Hence, we obtain the spectrum of the signal at the low-pass filter input as

$$Y_s(f) = X_s(f).P(f) \qquad (6.14)$$

where $P(f)$ is the Fourier Transform of the basic pulse $p(t)$. Then using Equation 6.10, the spectrum of the flat-top sampled signal becomes

$$Y(f) = \left[\sum_{k=-\infty}^{\infty} f_s X(f - kf_s) \right].P(f). \qquad (6.15)$$

From Chapter 2, we know that

$$P(f) = f\tau.\text{sinc}(f\tau)$$

which has zero values at $\pm 1/\tau$, $\pm 2/\tau$, ... The spectrum $Y_s(f)$ is illustrated in Figure 6.3(e). The output spectrum from a reconstruction low-pass filter is shown in Figure 6.3(f). We next examine the practical significance of these results.

6.2.2 Anti-aliasing Filter

Examination of Figure 6.3(d) reveals that if a signal spectrum extends to B and is sampled at a rate f_s higher than $2B$, then the spectrum of the sampled signal consists of periodic repetitions of $X(f)$. If the sampling rate is lower than $2B$, then the shifted components of $X(f)$ overlap and the spectrum of the sampled signal has only a distorted resemblance to the spectrum of the original signal.

This spectral overlap effect is known as *aliasing* and the distortion that results is called *aliasing distortion*. The sampling rate

$$f = 2B$$

below which aliasing occurs is known as the *Nyquist rate*.

Note that for a given analogue signal spectrum, any aliasing effects decrease as the spectral replicas move further apart and as the spectral components decrease more rapidly for frequencies near $f_s/2$. The former can be controlled by the choice of the sampling rate. The latter can be controlled by the use of a low-pass filter prior to sampling. The filter is referred to as an *anti-aliasing filter*.

For telephone speech signals with nominal maximum frequency $B = 3.3\,\text{kHz}$, it is usual to sample at $f_s = 8000$ samples/s. The gain-frequency response of a typical anti-aliasing filter is shown in Figure 6.4. It is a third-order elliptic low-pass filter which has an insertion loss of at least 20 dB for frequencies above 4.5 kHz.

6.2.3 Reconstruction Filter

By examination of Figures 6.3(e) and 6.3(f), it is apparent that flat-topped sampling gives rise to attenuation of high-frequency components in the low-pass signal. This effect is sometimes known as the *aperture effect*. The resultant high-frequency roll-off distortion is referred to as *reconstruction distortion*.

This distortion should be compensated by use of an equalizing reconstruction low-pass filter. Figure 6.5 illustrates the gain-frequency response of a typical reconstruction filter, which incorporates the necessary $(\sin x)/x$ correction factor.

Note that flat-top sampling has two significant advantages. Firstly, larger energy is contained in sample pulses of greater width resulting in increased output from the reconstruction filter. Secondly, the filtering provided by the

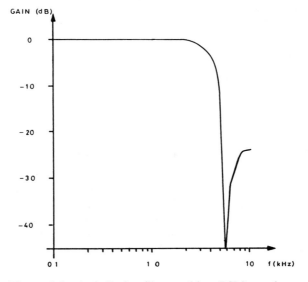

Figure 6.4   Anti-aliasing filter used in a PCM encoder.

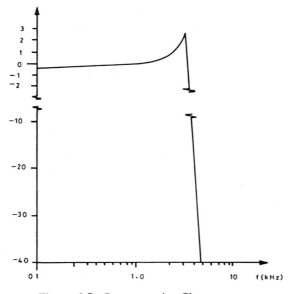

Figure 6.5   Reconstruction filter response.

aperture effect significantly attenuates frequencies above 4 kHz. This is beneficial in the reconstruction filtering.

Exercise 6.1

A sampled signal has the spectrum shown in Figure 6.6.

(1) What is the signal being sampled?
(2) What is the sampling rate?
(3) What is the sample pulse width?

Solution.

(1) A 1 kHz sinusoid, $x(t) = A \cos 2\pi.1000t$.
(2) The sampling rate f_s is given by the distance between the center frequencies of adjacent pairs of sidebands. It is $f_s = 5$kHz.
(3) The pulse width of the samples is obtained from the position of the first null in the $\mathrm{sinc}(f\tau)$ function which is at 50 kHz. Therefore, the sample pulse width is $\tau = 20$ μsec.

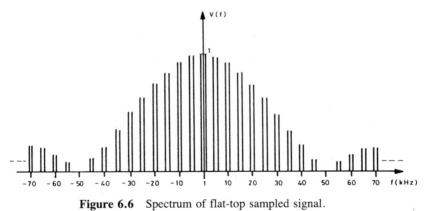

Figure 6.6   Spectrum of flat-top sampled signal.

6.2.4   Switched Capacitor Filter Methods

Modern speech coder/decoders (codecs) are designed as a single VLSI chip wherever possible. One difficulty is that the two low-pass filters must be of reasonably high order. Early filter designs were of the transversal digital filter type but needed many taps to get the necessary cut-off. Operational amplifier active filters have also been tried but they have accuracy and stability problems because of the difficulty in obtaining circuit elements with tight tolerances.

A break-through has occurred with the emergence of *switched-capacitor* filter techniques suitable for VLSI implementation. These use active filter configurations, but replace all resistors with switched-capacitor elements as illustrated in Figure 6.7.

The circuit shown in Figure 6.7(a) is a single-pole low-pass filter using conventional active filter techniques. It has the transfer function

$$\frac{v_2}{v_1} = \frac{1/R_1C_2}{s - 1/R_2C_2} . \tag{6.16}$$

The corresponding switched-capacitor filter form is shown in Figure 6.7(b) where the resistors in the previous circuit have been replaced by switched-capacitors. The switching function is performed by two metal-oxide semiconductor (MOS) switches driven by two-phase clocks ϕ_1 and ϕ_2 as shown in Figure 6.7(c). Equivalent representations of the switching operation are illustrated in Figures 6.7(d) and 6.7(e). They show that the capacitor is alternatively connected between the input voltage v_1 and the output.

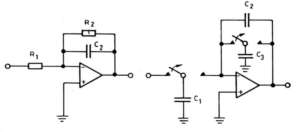

(a) CONVENTIONAL ACTIVE FILTER   (b) SWITCHED CAPACITOR FILTER

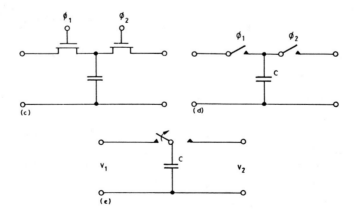

Figure 6.7 Switched-capacitor principles.

To see how the switched-capacitor circuit behaves as a resistor, consider the switch momentarily on the input terminal. The capacitor will be charged up to the voltage v_1. Typically this would take about 40 ns.

If the switch is now changed over to the output terminal, the capacitor will discharge to v_2 say, in one clock period. The charge transferred to the output terminal in one clock period $T_c = 1/f_c$ is

$$\Delta_q = C(v_1 - v_2).$$

If we assume that the clock period is small enough that the signals v_1 and v_2 hardly change during one clock period, then during this time interval, the current which flows is given to a good approximation by

$$i(t) = \frac{\Delta_q}{\Delta_t} = \frac{C(v_1 - v_2)}{T_c} . \tag{6.17}$$

Alternatively, we could have obtained the same current during the same time interval if we had placed an appropriate resistor R_c between the input and output terminals since then

$$i(t) = \frac{v_1 - v_2}{R_c} . \tag{6.18}$$

By equating Equations (6.17) and (6.18), we see that the size of the equivalent resistor to give the same current in the same time interval is

$$R = \frac{T_c}{C} = \frac{1}{f_c C} . \tag{6.19}$$

The resultant transfer function of the circuit in Figure 6.7(b) is

$$\frac{v_2}{v_1} = \frac{-f_c(C_1/C_2)}{s + f_c(C_3/C_2)} . \tag{6.20}$$

The values of the equivalent resistors in switched-capacitor circuits may be adjusted by adjusting the capacitor values or by varying the clock frequency. In audio filter circuits, the clock frequency is usually chosen to lie in the range 100 kHz to 2 MHz. The circuits require capacitor pairs with values such that their capacitance ratios are well controlled. Such pairs can be formed with accuracy in MOS technology so the filter characteristics can be realized with accuracy and stability.

6.3   PCM SYSTEM PERFORMANCE

A PCM encoder is shown in block schematic form in Figure 6.8(a). A typical codec structure in integrated circuit form is shown in Figure 6.8(b). For the purposes of this discussion, we ignore the multiplexing, signalling, and synchronization functions, which were discussed in earlier chapters. Here we wish to concentrate on the analysis of signal distortion caused in the encoding and decoding processes.

Let $x(t)$ be the output from the anti-aliasing low-pass filter. The low-pass analogue signal $x(t)$ is sampled 8000 times per second by a sampling circuit. We assume the encoder uses eight bits to represent each sample. That is, each sample is quantized and encoded into an 8-bit binary word $\{d_k\}$. This encoded 8-tuple is connected to a digital line, optical fibre, or radio system. Let the output regenerated word after transmission be $\{\tilde{d}_k\}$. The decoder converts the received sequence of 8-tuples into an analogue output.

The D/A converter section in Figure 6.8(b) converts the 8-bit binary words into a sequence of PAM samples. The decoded output signal $\tilde{x}(t)$ is reconstructed

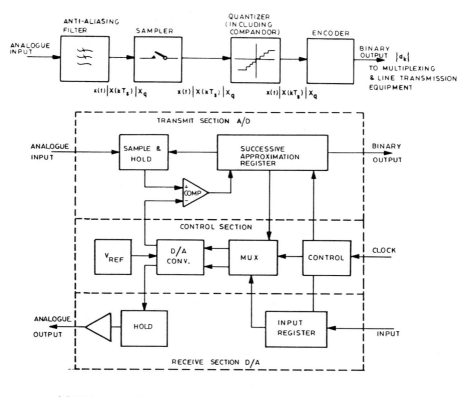

(b) TYPICAL PRACTICAL CODEC SCHEMATIC - INTEL 2911 SINGLE CHIP PCM CODEC (A LAW)

Figure 6.8 Pulse code modulation.

from these PAM samples using a low-pass filter made up of a sample-and-hold circuit on the codec chip and an external low-pass filter. The low-pass filter gain-frequency response must be designed to provide a $(\sin x)/x$ correction factor to compensate for the flat-top sampling errors as discussed in the previous section.

Distortion in the output speech waveform $\tilde{x}(t)$ can occur because of sampling/reconstruction errors, quantization effects, or channel errors during transmission. Sampling/reconstruction distortion was discussed in an earlier section. The other causes of distortion will be discussed next.

If we assume that the distortion that results from sampling and reconstruction is negligible, the reconstructed signal $\tilde{x}(t)$ can be expressed in terms of the input $x(t)$ as

$$\tilde{x}(t) = k\,x(t) + n_q(t) + n_o(t) \qquad (6.21)$$

where $\qquad k =$ a gain constant

$n_q(t) =$ distortion resulting from quantization

$n_o(t) =$ distortion resulting from regenerator errors in the transmission system.

The distortion functions $n_q(t)$ and $n_o(t)$ are commonly referred to as ''noise'' components. Next we consider each of these components in more detail.

6.3.1   Quantization Noise

The quantization noise component $n_q(t)$ arises because of the limits in accuracy involved in the process of converting signal samples into a finite number of n-bit binary words. A distinct n-bit word can be associated with one of only 2^n values.

A sampler is assumed to sample the analogue waveform $x(t)$ to produce the sample values $X(kT_s)$ for $k = 0, 1, \ldots$. Note that upper-case symbols are used in this section to represent the variables. Lower-case symbols are used to specify particular values taken by a variable.

Figure 6.9(a) illustrates the quantizing process. The quantizer converts each of the input sample values $X(kT_s)$ to output values $X_q(kT_s)$. Each output value occupies one of Q allowable output levels $y_1, y_2, \ldots y_Q$ using some predetermined rule

$$X_q(kT_s) = y_i \qquad \text{if} \qquad x_{i-1} \le X(kT_s) < x_i \qquad (6.22)$$

The values $x_1, x_2, \ldots, x_{Q-1}$ represent quantizer region boundaries, as illustrated in Figure 6.9(a).

The quantization process can be modelled as the addition of a quantizing error component

$$e(kT_s) = X_q(kT_s) - X(kT_s) \qquad (6.23)$$

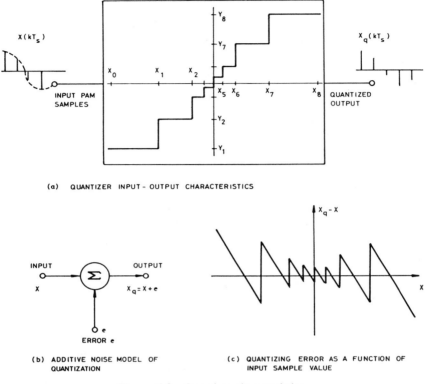

(a)   QUANTIZER INPUT - OUTPUT CHARACTERISTICS

(b) ADDITIVE NOISE MODEL OF
    QUANTIZATION

(c)   QUANTIZING ERROR AS A FUNCTION OF
      INPUT SAMPLE VALUE

Figure 6.9   Quantizer characteristics.

to the input as illustrated in Figure 6.9(b) where we have used the simpler representation $X = X(kT_s)$ and $X_q = X_q(kT_s)$.

The performance of a quantizing scheme is commonly measured in terms of the *signal to quantizing noise power ratio* defined as

$$\text{SNR}_Q = 10 \log_{10} \frac{\sigma^2}{E\{(X_q - X)^2\}} \tag{6.24}$$

where σ^2 is the variance of the input signal and the symbol $E\{.\}$ represents ensemble averaging.

Uniform Quantizers

Of special interest is the *uniform quantizer* where the decision levels are equally spaced. Then the range of the continuous amplitude values of the samples $X(kT_s)$ is divided into Q intervals of equal length as illustrated in Figure 6.10.

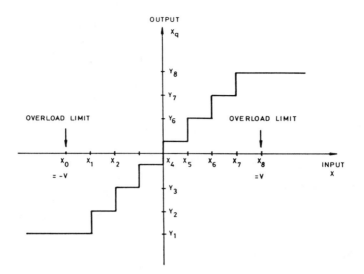

Figure 6.10   Uniform quantizer ($Q = 8$ levels).

In practice, to fully specify a uniform quantizer, it is necessary to specify overload levels $+V$ and $-V$ which determine the range of input voltage values to be equally subdivided. Then the step size or interval length Δ is

$$\Delta = 2V/Q. \tag{6.25}$$

The quantization noise is classified as either:

(1) *granular* noise when the input sample X lies within the interval
$x_o \leq X \leq x_Q$,
(2) *overload* noise when X lies outside that interval, or
(3) *idling* noise when the input is zero.

It is often convenient to artificially model quantization noise, when the input is nonzero, as the sum of granular and overload noise as if they were two separate noise sources.

For a uniform quantizer, the choice of suitable overload levels

$$x_o = -V \text{ and } x_Q = +V$$

should be made so as to avoid significant overload noise. To do this, it is convenient to define a *loading factor* as the ratio of overload limit to input signal rms value, that is

$$\text{Loading Factor} = V/\sigma. \tag{6.26}$$

Next, a suitable loading value factor must be chosen for the quantizer. A common choice is

$$V/\sigma = 4$$

which is referred to as *four-sigma loading*. Then the step size is

$$\Delta = 8\sigma/Q. \qquad (6.27)$$

Exercise 6.2

Consider a uniform 8-step quantizer with overload values $V = \pm 0.8$ volts. A periodic triangular waveform of peak amplitude V_p volts is the input signal $x(t)$ to a sampler and quantizer. The sampler samples the signal 8 times each period.

Sketch the quantization noise as a function of time for one period of the input signal for two cases, $V_p = 0.4$, and $V_p = 1.0$ volts, respectively.

Solution. The results are obtained graphically and shown in Figure 6.11. Note that for $V_p = 0.4$ only granular noise effects are evident as shown in Figure 6.11(c).

Exercise 6.3

Suppose that the quantizer input signal samples X are modelled as Gaussian random variables with zero mean and variance σ^2. If four-sigma loading is used, find the probability that overload occurs.

Solution. Overload will occur if samples of the input signal fall outside the overload limits -4σ to 4σ. This occurs with probability

$$P(\text{overload}) = 2 \int_{4\sigma}^{\infty} \frac{1}{\sqrt{2\pi}\,\sigma}\, e^{-x^2/2\sigma^2}\, dx.$$

Letting $t = x/\sigma$, the lower integral limit becomes 4 and we obtain

$$P(\text{overload}) = 2\, Q(4) \simeq 0.6 \times 10^{-4}$$

where we have used values for the $Q(.)$ function from Figure 4.24 in Chapter 4.

The noise power associated with uniform quantization can be found using the mean square value

$$N = E\{(X_q - X)^2\}$$

and using Equation (6.23) this can be evaluated as

$$N = \int e^2\, p(e)\, de \qquad (6.28)$$

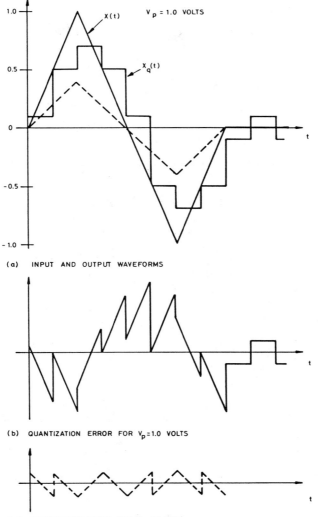

(a) INPUT AND OUTPUT WAVEFORMS

(b) QUANTIZATION ERROR FOR $V_p = 1.0$ VOLTS

(c) QUANTIZATION ERROR FOR $V_p = 0.4$ VOLTS

Figure 6.11 Quantization error example (granular noise only).

where $p(e)$ is the probability density function of the quantizing error. Let us assume that overload noise occurrences are negligible.

For typical quantization noise of the granular type (for example Figure 6.11(c) and lacking information to the contrary, it is usual to assume that e is uniformly distributed over the possible range $-\Delta/2$ to $\Delta/2$. It follows that, within that interval, $p(e) = 1/\Delta$.

Then the mean-square value of the quantizing noise becomes

$$N = \int_{-\Delta/2}^{\Delta/2} e^2 . (1/\Delta)de \tag{6.29}$$

and hence, we obtain

$$N = \frac{\Delta^2}{12}. \tag{6.30}$$

Then using Equations (6.24) and (6.30), we obtain

$$\text{SNR}_Q = 10 \log \frac{\sigma^2}{\Delta^2/12}. \tag{6.31}$$

Consider the number of quantizer levels to be

$$Q = 2^n, \quad Q \gg 1.$$

Then since $\Delta = 8\sigma/Q$ from Equation (6.27), we obtain for an n-bit quantizer that the signal to quantization noise ratio is

$$\text{SNR}_Q = 6n - 7.3 \text{ (dB)}. \tag{6.32}$$

That is, the SNR increases linearly with the number of bits per sample (n). Changing the loading factor modifies the 7.3 term, but does not alter the rate of increase of SNR with n.

Let us review the significance of these results. The above SNR value is the ratio of the signal power (normalized in 1 ohm) at the quantizer input to the quantizing noise power. Consider the overall PCM system illustrated in Figure 6.3(a). If the transmission system is error-free, then the successive sample values represented in the signal $y_s(t)$ at the transmission system output will be identical to the sample values $X_q(kT_s)$ at the quantizer output in Figure 6.8(a). That is, Equation (6.32) gives us the signal-to-noise ratio at the transmission system output (Figure 6.3(a)). This is for the case where it is assumed that the overload noise and transmission errors are negligible.

Now the quantized signal $y_s(t)$ is passed through a reconstruction low-pass filter to obtain the actual system output $\tilde{x}(t)$. The smoothing effect of the low-pass filter has been found, in practice, to result in an improvement in the signal to quantization noise ratio by about 6 dB.

What about the effects of channel errors on the output signal to noise ratio? Consider a PCM system using uniform quantization. If the PCM signal is transmitted over a single regenerator section without repeaters, we can estimate and sketch the output signal to noise ratio as a function of regenerator input signal to noise ratio.

Note, from Chapter 4, that when the regenerator input signal to noise ratio is decreased below approximately 15 dB, channel errors increase rapidly and will become the dominant source of output noise. When $S_i/N_i \gg 15$ dB, channel errors are negligible and the system is quantization noise limited. Figure 6.12 summarizes the signal to noise ratio performance for the cases where $n = 7$ and $n = 8$, assuming four-sigma loading.

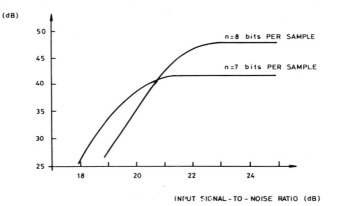

Figure 6.12   PCM system performance: Uniform companding (signal input probability density function assumed uniform).

6.3.2   Companding Techniques

In considering the performance of a quantizer, examining only the signal to quantization noise ratio SNR_Q for a fixed input signal level σ is not sufficient. The dynamic range must also be examined. Varying the input signal level for a given uniform quantizer will vary the value of SNR_Q. This occurs because for low input signal values, the quantization is too coarse, and for excessively large input signal values, overload noise becomes important.

The value of SNR_Q will depend on the input signal level and the probability distribution of the input signal. In Figure 6.13, the dependence of signal to noise ratio on input power level is shown for a uniform quantizer with $n = 7$ and $n = 8$ bits, respectively. These signal-to-quantization noise curves are derived from Bylanski (11) assuming that the input signal probability density function is the Laplacian density

$$p(x) = \frac{1}{\lambda} \exp \frac{(-2|x|)}{\lambda} \tag{6.33}$$

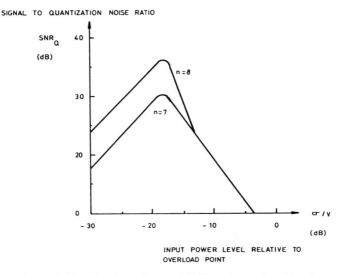

Figure 6.13 The dependence of SNR on input signal level.

where the mean square value of the signal is

$$\sigma^2 = \frac{\lambda^2}{2}.$$

The Laplacian probability density model is sometimes used to approximate speech characteristics as discussed in Chapter 4.

It is noted, in passing, that another probability model for speech is the Gamma density

$$p(x) = \{\sqrt{k/(4|x|)}\}\exp(-k|x|). \tag{6.34}$$

It is a more complex model but may be a better approximation. See for example Paez (1) for further details.

The region to the left of the maxima in Figure 6.13 represents the fall-off in SNR because of reduced input signal levels, that is, reduced loading. To the right of the maxima, the fall-off is due to overload noise caused by large input levels.

If a minimum acceptable SNR_Q level is 30 dB say, it can be seen from Figure 6.13 that the uniform quantizer has a useful dynamic range of only about 2 dB for $n = 7$ and 10 dB for $n = 8$. This could not provide acceptable quality for speech transmission in public networks. For a linear quantizer to provide acceptable speech quality, at least 11 bits per sample are required. This would obviously require a proportional increase in bit rates.

The use of nonuniform quantization can considerably improve the useful dynamic range of operation by providing a constant SNR_Q over a much larger dynamic range of input signals. In the nonuniform quantizer, more quantizer levels are allocated to the smaller input signal amplitudes, which generally have higher probability. Fewer levels are allocated to the less frequently occurring higher amplitudes. This technique is called *companding*, a term combining "compressing" and "expanding."

It is possible to model nonuniform quantization, as shown in Figure 6.14. The model consists of a memoryless nonlinearity $F(x)$ representing the compressor. This is followed by a uniform quantizer. The output values are then applied to the inverse nonlinearity $F^{-1}(x)$ referred to as the expander. The function $F(x)$ is known as the companding law.

Despite a great deal of international effort, a single companding law has not been accepted universally. Two different companding laws have been accepted by CCITT, (Rec.G.711) namely:

(1) the A-law (adopted in Europe, Asia, and Australia)
(2) the North American μ-law.

Both are segmented piecewise-linear approximations to continuous functions, given for a range of ± 1 as
A-law (Europe, Asia, Australia)

$$F_A(x) = \begin{cases} \text{sgn}(x)\ \dfrac{A|x|}{1 + \log_e A} & \text{for } 0 \le x \le 1/A \\[3mm] \text{sgn}(x)\ \dfrac{1 + \log_e A|x|}{1 + \log_e A} & \text{for } 1/A \le x \le 1 \end{cases} \qquad (6.35)$$

μ-law (N. America)

$$F_\mu(x) = \text{sgn}(x)\ \frac{\log_e(1 + \mu|x|)}{\log_e(1 + \mu)} \quad \text{for } -1 \le x \le 1. \qquad (6.36)$$

A value of $\mu = 255$ is generally used in practice. Some μ-law systems were in existence before the international standardization processes were completed

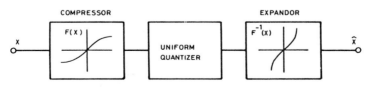

Figure 6.14   Model of nonuniform quantization.

and were a de facto standard which needed to be accommodated. It should be noted that CCITT Rec.G.711 states that ''Digital paths between countries which have adopted different encoding laws should carry signals in accordance with A-law.''

The CCITT quantizer compression recommendations envisage segmented versions of these laws, that is, piece-wise linear approximations to the smooth functions illustrated in Figure 6.14. A segmented A-law is illustrated in Figure 6.15(a).

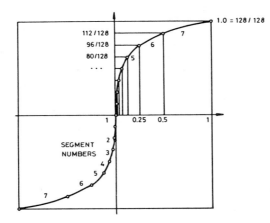

(a) A-LAW COMPANDING CHARACTERISTIC

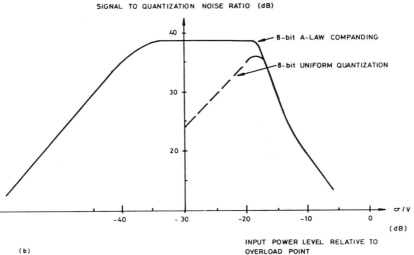

(b)

Figure 6.15 Companding.

The A-law standard uses $A = 87.6$ and is defined for positive input values as shown in Table 6.1. There are seven segments for positive values and 13 segments in all (since segment 0 is common to both positive and negative values). Each code word consists of eight binary bits of the form

$$\{d_k\} = (PXYZABCD)$$

where

P = the polarity digit

XYZ = the segment digits (indicating the segment in which the input value lies)

$ABCD$ = the linear digits (indicating the amplitude value in the segment).

The coder input range (overload limit) is considered as

$$V = 4096 \text{ units}$$

so that the smallest quantum interval (used for very small inputs) is $V/4096$ units.

The measured A-law quantizer signal to noise ratio for a Gaussian input is shown in Figure 6.15(b). It is apparent from this diagram that companding leads to a much larger useful dynamic range of input signal values. For an SNR_Q minimum level of 30 dB, the dynamic range has been increased from 10 dB (for the 8-digit uniform quantizer) to approximately 45 dB.

Table 6.1
Definition of segmented PCM A-law

Segment No. S $(S = 4X + 2Y + Z)$	Coder input range	PCM Output code word	Quantum interval
0	$0 \quad -V/128$	P000ABCD	2
1	$V/128-V/64$	P001ABCD	2
2	$V/64 \quad -V/32$	P010ABCD	4
3	$V/32 \quad -V/16$	P011ABCD	8
4	$V/16 \quad -V/8$	P100ABCD	16
5	$V/8 \quad -V/4$	P101ABCD	32
6	$V/4 \quad -V/2$	P110ABCD	64
7	$V/2 \quad -V$	P111ABCD	128

6.3.3   CCITT Standards for Quantization Noise

The international telecommunications standards recommended by the CCITT (Rec.G.721) for quantization distortion limits are based on a hypothetical sequence of PCM links known as the *International Reference Circuit*. As discussed in Chapter 5, this circuit is intended to represent the longest expected worldwide circuit and is assumed to contain 14 coding-decoding processes in tandem. Note that this involves conversion from digital-to-analogue signals and back to digital signals again at each stage.

For this reference circuit, the minimum overall signal to quantization noise ratio is recommended to be

$$\text{Min. SNR}_Q = 22 \text{ dB}$$

If we assume that the quantization noise components contributed in each of the 14 coder-decoder segments are independent, then we can add the 14 noise power values to obtain the overall quantization noise power. If each segment is assumed to generate the same amount of quantization noise power, then the minimum SNR for each segment must be 14 times that of the overall circuit. That is

$$\text{Per segment min. SNR}_Q = 22 + 10 \log 14 = 34 \text{ dB}.$$

Figure 6.16 shows the recommended CCITT (Rec.G.712) specification values for a single segment as a function of input signal level. Also shown are the theoretical signal to quantizing noise ratios for a Gaussian noise input for the A-law and μ-law compandors. See Schweizer (2).

Quantizing Distortion (QD) Units

The CCITT (Rec.G.113) has also recommended a method of specifying the maximum contributions from various sources of quantization noise in one PCM coder-decoder segment. This is expressed in terms of *Quantization Distortion*

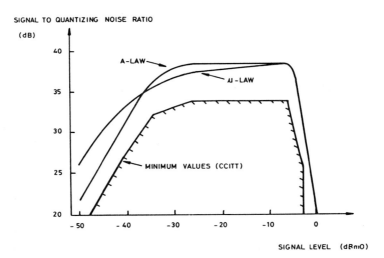

Figure 6.16 CCITT standards for SNR and A-law and μ-law performance for 8 bit encoding.

Units, also referred to as QD units. One QD unit is defined as the quantizing distortion power of one 8-bit PCM encoder-decoder process.

The "5 + 4 + 5 QD unit" Rule

For a complete international connection (14 PCM segments), the CCITT recommends that there should be a maximum of 14 QD units of quantizing noise power, five allocated to each national part of the circuit and four allocated to the international part (between gateway exchanges). See Schweizer (2) for details.

Furthermore, contributions by various sources are specified in terms of the QD count. This is to include such contributions as those caused by the need for conversion from A-law to μ-law at the boundaries of two national systems using different standards. It is recommended that these be determined as shown in Table 6.2.

Table 6.2
CCITT recommended contributions to the QD count.

Process or system	QD Units
Standard 8-bit PCM, A or μ-law (encoder + decoder)	1
7-bit PCM encoder	4
A-μ or μ-a code conversion	1
μ-A-μ code conversion (tandem, within 1 digital path	0.5
Digital attenuation, 1 to 8 dB (A or μ-law)	1
32 kbit/s adaptive differential PCM (ADPCM)	3–4

6.4 CODING TECHNIQUES FOR REDUCED BIT RATES

PCM encoding procedures can provide good quality speech transmission at 64 kbit/s. A great deal of research has been directed in recent years to the search for more efficient waveform encoding methods which can reduce the required bit rates. These techniques are sometimes referred to as *data compression* or *bit rate reduction* techniques.

6.4.1 Principles of Data Compression

Signals such as speech and video exhibit considerable redundancy, which makes them amenable to data compression. In speech, the redundancy results from the physical mechanism of the human vocal tract and the inherent structure of language. Also, our ability to perceive speech and other sounds is constrained in dynamic range and bandwidth because of the physical mechanism of the human ear.

The process of the production of speech can be represented by the electrical model shown in Figure 6.17. As represented by this model, speech is generated by a sound source excitation that is either *pulse-like*, during "voiced" regions of speech, when the vocal cords are vibrating, or *noise-like*, during "unvoiced" regions of speech because of turbulence at constriction points in the vocal tract. These pulse-like or noise-like excitations are then filtered by the vocal tract, which behaves as a time-varying acoustic filter.

When speech signals are sampled, there is a high sample-to-sample correlation. This is because when speech waveforms are examined over a relatively short period of time (of the order of 20–50 ms), the signals are statistically stationary and have a well-defined short-time frequency spectrum. By contrast, when viewed over a relatively long period of time (of the order of 1 sec), the variations of amplitude in the voiced, unvoiced, and silent periods result in highly nonstationary signals. The short-time correlations in the speech signals can be used for more efficient encoder techniques.

Likewise, in video (picture) information, appreciable portions of the signal represent background information containing very little tonal variations. If we transmitted this signal using conventional PCM, the sample codewords would not change appreciably from code word to code word.

Efficient digital coding techniques also take into account the perception processes of hearing and seeing. For speech, the ear behaves like a filter bank, each channel of which appears to have its own automatic gain control and dynamic range limitations. See Flanagan (3) for details.

PCM systems with companding can take advantage of the long-term amplitude nonstationarity by using variable quantizer step sizes to take account of the high-crest factor (ratio of peak to rms value) of speech. This provides good quality at 64 kbit/s.

As further advantage is taken of short-term time correlation or dynamic spectral characteristics, good quality can be achieved at bit rates in the range of 24 to 32 kbit/s. By including the voice production properties of pitch and spectral shaping, acceptable performance can be achieved at bit rates down to 9.6 kbit/s.

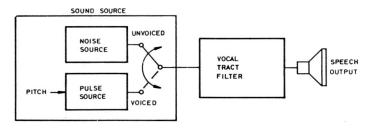

Figure 6.17   Electrical model of speech production.

Waveform coding methods for speech can be subdivided into two types:

(1) *time domain methods*, in which the speech signal is sampled and then the samples are processed taking into account the correlation and dynamic spectral characteristics. Time domain waveform coding techniques include

> Pulse code modulation (PCM)
> Adaptive pulse code modulation (APCM)
> Differential PCM (DPCM)
> Delta modulation (DM)
> Adaptive delta modulation (ADM)
> Adaptive differential PCM (ADPCM)
> Adaptive predictive coding (APC)

(2) *frequency domain methods*, in which the speech signal is subdivided into a set of separate frequency components and these components are independently coded. Frequency domain waveform coding techniques include the technique referred to as Subband coding (SBC).

By contrast, as discussed in Section 6.1 above, vocoding methods have been developed as an alternative to waveform coding. Vocoding methods use source modelling. They attempt to approximate the speech production model of Figure 6.17. Speech parameters defining the vocal tract filter, pitch, and voiced/unvoiced/silence decisions are encoded in the transmitter. In the receiver they are used to drive the electrical model of Figure 6.17. Vocoder methods include

(1) linear predictive coding (LPC)
(2) cepstrum modelling

A further class of speech coding procedures use *hybrid* or *parametric methods*. These combine the features of both waveform coding and vocoding in an attempt to fill the performance gap between these two methods. Hybrid methods include

(1) voice excited vocoding
(2) harmonic scaling

Historically, vocoders were among the earliest types of speech compression techniques. In general, the speech quality of vocoders is not as high as that of waveform coding techniques. However, they do permit intelligible encoding at bit rates as low as 1 to 2 kbit/s, although the resultant speech has an unnatural "machine-like" quality and significant processing delays may occur. Linear predictive coders operating at bit rates in the range 2.4 to 4.8 kbit/s are of particular interest in military communication systems.

The hybrid or parametric techniques have resulted in encoders working in the range of 4 to 10 kbit/s. In one such method, referred to as voice-excited vocoding, a portion of the speech band, say 0–800 Hz, is encoded using waveform coding methods. The remainder of the speech band, 800–4000 Hz, is vocoded.

Further details on vocoder and hybrid coding techniques can be found in Crochiere (4) and Jayant (12). In the remainder of this chapter, we examine some *waveform encoding* techniques that appear likely to have application in public networks using digital speech transmission.

6.4.2   Adaptive Pulse Code Modulation (APCM)

In conventional PCM systems, companding techniques involve nonuniform quantization in an attempt to match the quantizer step size to the amplitude probability density function of the input signal. Adaptive pulse code modulation (APCM) techniques provide an alternative approach in that they use uniform quantizer intervals or step sizes. However, the size of each step may vary as a function of time with the rms level of the signal. In this way, the dynamic characteristics of amplitude nonstationarity of speech may be more efficiently tracked.

The step size can be adjusted in a self-adapting manner based on observation of past code words. Hence, if d_i represents the step size at time iT_s, then the next value d_{i+1} may be larger or smaller depending on the value of weighted sums of the previous quantizer outputs. That is

$$d_{i+1} = d_i f(X_q) \qquad (6.37)$$

where $f(X_q)$ represents some function of the previous quantized values X_q.

At first thought, it might seem that the same effect could be achieved equally well by using a uniform quantizer with fixed step size, preceded by an amplifier with automatic gain control. While this could be used to produce the required transmit code sequence, it does not provide information to the receiver to indicate how its complementary automatic gain control amplifier should adapt. With an APCM system, as long as there are no channel errors, the correct step size value can be determined at the receiver by observations of previous quantized samples and use of the same function $f(X_q)$ as at the transmitter.

6.4.3   Differential Pulse Code Modulation (DPCM)

Differential coding methods such as differential pulse code modulation (DPCM) utilize the properties of sample-to-sample correlation in speech (owing to the low-pass characteristic of the speech spectrum) and the dynamic spectral characteristics (owing to the vocal tract structure).

In a DPCM system, illustrated in Figure 6.18(a), the basic aim is to transmit the difference between the present sample value $X(kT_s)$, and a predicted value $Z(kT_s)$ based on past samples. A linear transversal filter is used to estimate or

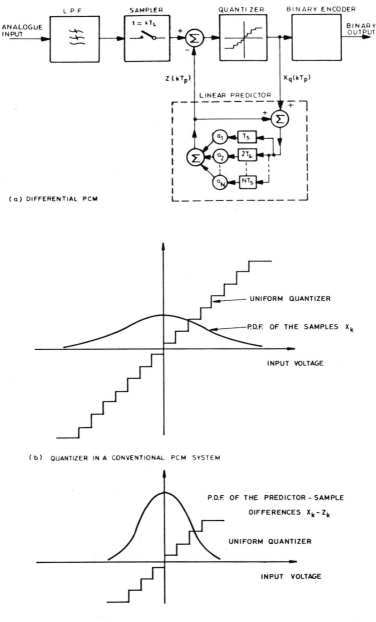

Figure 6.18 Reduction of the rms value at the quantizer input for differential PCM.

predict future sample values $Z(kT_s)$ based on knowledge of previous quantizer outputs and the characteristics of the vocal speech generating system. The filter coefficients $a_1, a_2, \ldots a_N$ are fixed. They are chosen to model the vocal tract filter. The predicted value at the filter output is

$$Z(kT_s) = a_1 X_q((k - 1)T_s) + a_2 X_q((k - 2)T_s)$$

$$+ \ldots + a_N X_q((k - N)T_s). \quad (6.38)$$

In practice, only one nonzero predictor coefficient is often used (that is, $N = 1$). A value of $a_1 = 0.85$ is reported to provide good results. See (8) for details.

If the predictor makes a good estimate of the sample value, the differences

$$X(kT_s) - Z(kT_s)$$

will have much smaller rms level (variance) than the rms level of the input signal. As illustrated in Figure 6.19, this permits a smaller quantizer step size to be used, which leads to reduced quantization noise. Alternatively, the same level of quantization noise can be achieved with fewer steps (fewer binary digits per code word). Thus, a speech signal that has been quantized to 256 levels (eight bits per code word) using conventional PCM may be transmitted with comparable quality using 16 quantization levels (four bits per code word). This reduces the required transmission bit rate by a factor of 2.

DPCM systems with fixed predictors can provide considerable improvement in signal-to-quantization noise ratio over direct quantization as in PCM. The greatest improvement is reported in going from no prediction to first-order prediction ($N = 1$ in Figure 6.18) with smaller additional gains resulting from increasing the prediction order up to 4 or 5. After that, little additional gain in SNR performance is obtained.

6.4.4   Delta Modulation (DM)

Delta modulation is a special case of DPCM. To see this, consider the block schematic for a DPCM system in Figure 6.18(a). If only one delay element is used in the predictor filter with $a_1 = 1$ and if there are just two quantization levels, then the system becomes a Delta modulator as illustrated by the discrete time model shown in Figure 6.19(a). In this case, the sample-to-sample difference is quantized, the result being one of only two levels which becomes the binary output.

In Figure 6.19(b) an equivalent hardware implementation is shown. This illustrates the simplicity of a delta modulator. The demodulator consists of even less components, namely an integrator and a low-pass filter as shown in Figure 6.20.

In the delta modulator shown in Figure 6.19(b), the input signal $x(t)$ is compared with the predicted or estimated signal $\tilde{x}(t)$. The comparator output is

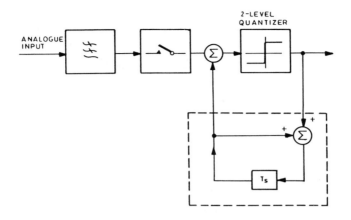

(a) DISCRETE TIME MODEL

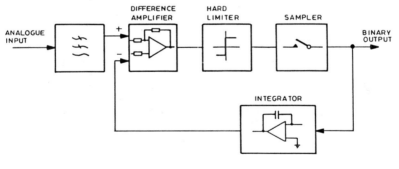

(b) HARDWARE IMPLEMENTATION

Figure 6.19 Delta modulation.

$$x_L(t) = \text{sgn}\,[x(t) - \tilde{x}(t)] \tag{6.39}$$

where

$$\text{sgn}[x(t) - \tilde{x}(t)] = \begin{cases} +1 \text{ for } x(t) > \tilde{x}(t) \\ -1 \text{ for } x(t) \le \tilde{x}(t). \end{cases}$$

The comparator output $x_L(t)$ is sampled $f_s = 1/T_s$ times per second to produce the output sequence of binary pulses. Note that the sample rate in a DM system is usually 32 or 64 kbit/s.

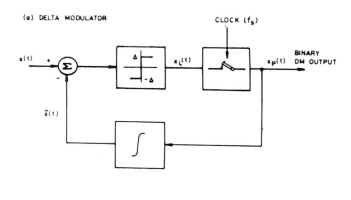

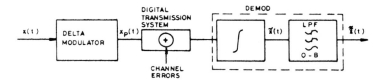

Figure 6.20   A delta modulation system.

In the modulator, the binary output $x_p(t)$ is fed back through the integrator, which forms a staircase approximation $\tilde{x}(t)$ to the original signal. This is illustrated in Figure 6.21. Hence, the comparator input is the difference between the input signal and the approximation to the input formed out of all previous output information.

The output binary sequence can be written

$$x_p(t) = \sum_{k=-\infty}^{\infty} \Delta \ \text{sgn} \ [x(t) - \tilde{x}(t)]\delta(t - kT_s). \tag{6.40}$$

Likewise, the demodulator must also form an approximation to the original signal. It can do this using an integrator in the same way that the modulator uses an integrator. The resultant demodulator is shown in Figure 6.20. In a practical demodulator, a low-pass filter will perform a measure of integration so the integrator circuit may be omitted.

The staircase approximation $\tilde{x}(t)$ to the original signal is constrained to move up or down by only one quantization level at each sampling instant. This can lead to some problems, as illustrated in Figure 6.21. They can be classified as

(1) *start-up*—during the initial intervals when, if $x(t)$ is greater than $\tilde{x}(t)$, the staircase approximation may differ significantly from the input signal until $\tilde{x}(t)$ becomes greater than $x(t)$

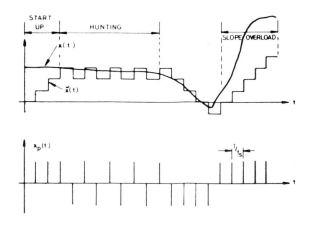

Figure 6.21   Delta modulation waveforms.

(2) *hunting*—when the input $x(t)$ remains constant but $\tilde{x}(t)$ consists of alternatively positive and negative steps, giving rise to what is known as *idling
noise*

(3) *slope overload*—if the rate of change of the input $x(t)$ exceeds the maximum
rate of change of the staircase approximation $\tilde{x}(t)$.

The key to the effective design of a DM system is the intelligent choice of the
two parameters, *step size* and *sampling rate*. To see how these affect the performance, we will next find an expression for the signal-to-quantization noise
ratio for a DM system.

Let $\bar{x}(t)$ represent the integrator output in the demodulator in Figure 6.20.
Neglecting channel errors, $\tilde{x}(t)$ must be the same as the staircase approximation
in the modulator. That is

$$\bar{x}(t) = \tilde{x}(t). \tag{6.41}$$

Now the quantization noise is the difference between the input $x(t)$ and the output
$\bar{x}(t)$, which we can write

$$e_q(t) = x(t) - \tilde{x}(t) \tag{6.42}$$

where

$$|e_q(t)| \le \Delta \tag{6.43}$$

if we assume that slope overload is avoided. We wish to find the effect of this
noise at the output of the demodulator low-pass filter. If the filter is assumed to
be an ideal low-pass filter with bandwidth B, just sufficient to pass all the
frequency components in the input signal, then its output can be written as

$$\overset{\approx}{x}(t) = x(t) + n_q(t) \tag{6.44}$$

where $n_q(t)$ is the quantization noise at the output of the low-pass filter.

Next, we assume that $e_q(t)$ is equally likely to have any value in the range $-\Delta$ to $+\Delta$, (that is, it has a uniform probability density function). Then, $e_q(t)$ has the mean square value

$$E\{e_q^2(t)\} = \int_{-\Delta}^{\Delta} \frac{1}{2\Delta} e^2 \, de = \frac{\Delta^2}{3}.$$

This represents the noise at the input to the low-pass filter. To obtain the output value, we note that it has been found experimentally that the power spectrum of the waveform $e_q(t)$ is constant over the frequency interval 0 to f_s and can be written in one-sided form as

$$S_e(f) = \begin{cases} K, & 0 < f < f_s \\ 0, & \text{otherwise.} \end{cases}$$

It follows that the spectral density is

$$K = \frac{\Delta^2}{3f_s}$$

and the mean square output of the low-pass filter is

$$E\{n_q^2(t)\} = \frac{\Delta^2 B}{3f_s}. \tag{6.45}$$

In order to calculate the output SNR_Q, we assume a sinusoidal modulation signal at the maximum signal frequency B (the worst case). Then the signal output may be written

$$x(t) = A \cos 2\pi B t \tag{6.46}$$

which has mean square value $A^2/2$. Then, from Equation (6.45) the signal to quantization noise ratio becomes

$$\text{SNR}_Q = \frac{3A^2}{2\Delta^2} \cdot \frac{f_s}{B}. \tag{6.47}$$

Note that the sampling rate f_s must be chosen in relation to the parameters A and B in such a way that slope overload is avoided. This gives us a further constraint on the achievable SNR_Q value. The maximum slope of the input signal is from Equation (6.46)

$$\max\left[\frac{dx(t)}{dt}\right] = \max[-2\pi B A \sin 2\pi B t] = 2\pi B A. \tag{6.48}$$

The maximum slope of the staircase waveform $x(t)$ is Δf_s so to avoid slope overload, we must have

$$\Delta f_s \geq 2\pi BA. \tag{6.49}$$

Hence, the maximum peak value A of the signal that avoids slope overload is

$$A = \frac{\Delta f_s}{2\pi B}. \tag{6.50}$$

On substitution in Equation (6.47), we obtain the signal to quantization noise as

$$\text{SNR}_Q = \frac{3f_s^3}{8\pi^2 B^3}. \tag{6.51}$$

It is interesting to compare the performance of DM and PCM systems. For equitable comparison, we assume that both systems use the same transmission bit rates. For PCM with n-bit code words and 2B samples per second, the sampling rate is

$$r_s = 2nB \tag{6.52}$$

and on substitution in Equation (6.51) we obtain for DM

$$\text{SNR}_Q = \frac{3(2n)^3}{8\pi^2}.$$

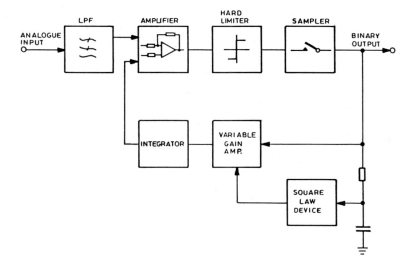

Figure 6.22 Adaptive delta modulation.

When expressed in dB this becomes

$$10 \log_{10} \text{SNR}_Q = 30 \log_{10} n - 5.3. \quad \text{(dB)} \qquad (6.53)$$

We can compare this with the result for PCM in Equation (6.32), reproduced here for convenience

$$10 \log_{10} \text{SNR}_Q = 6n - 7.3. \quad \text{(dB)}$$

For $n = 8$, we obtain SNR values of 21.9 dB for the DM system and 41.7 dB for the PCM system. Clearly, the performance of DM systems as measured by signal to quantization noise ratio, falls below the performance of PCM systems using the same bit rate and bandwidth.

The use of adaptively varied step sizes can greatly improve the performance of delta modulation systems. A method of doing this is illustrated in Figure 6.22. This is referred to as Adaptive Delta Modulation (ADM). The performance of ADM for speech is competitive with PCM. For details, see Jayant (5).

6.4.5   Adaptive Differential Pulse Code Modulation (ADPCM)

The concept of differential coding can be extended to be adaptive to the dynamic variations in speech levels. This is achieved by using differential encoding in conjunction with adaptive step-size methods, as discussed in Section 6.4.2. This is known as Adaptive Differential Pulse Code Modulation (ADPCM). Differential encoding improves the signal to quantization noise performance of the system. The addition of adaptive quantization improves the dynamic range as well as the SNR_Q.

Figure 6.23(a) illustrates an ADPCM system with adaptive quantization of the feedback type. Figure 6.23(b) depicts an ADPCM with feed-forward adaptive quantization.

Systems using feed-forward adaption require the current step size to be transmitted with the code word. Recall from Section 6.4.2 that when feedback adaptive methods are used, the step size does not have to be transmitted because the receiver can determine the required step size from the incoming sequence of code words. However, the quality of the reconstructed output will be more sensitive to errors in transmission.

ADPCM techniques offer the possibility of good telephone quality speech at a bit rate of 32 kbit/s. This means that ADPCM can result in a 50 percent bit-rate reduction over the conventional PCM 64 kbit/s rate. The standardization of an ADPCM algorithm for 32 kbit/s telephone transmission is currently being undertaken within the CCITT. In principle, the encoder should be designed to be signal independent, and therefore, able to handle not only speech but also a variety of signals such as music, tones, and voiceband data. These constraints

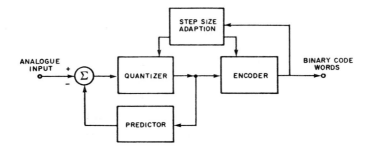

(a) ADPCM with feed-back adaptive quantization

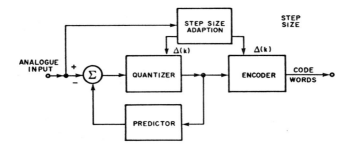

(b) ADPCM with feed-forward adaptive quantization

Figure 6.23   ADPCM encoding schemes.

are likely to increase the complexity of the algorithms to be used in ADPCM systems.

Figure 6.24 shows the block diagram of a typical ADPCM codec that has been found to be robust to channel errors and provide good quality speech. The predictor feedback loop contains a first-order fixed predictor with input sample value dependent on step size and predictor coefficient $\alpha = 0.85$.

Feedback adaption of the linear quantizer step size is used. After each speech sample is encoded, the quantizer step size changes. For a 32 kbit/s coder with an 8 kHz sampling rate, the number of quantizer steps is

$$2N = 2^4 = 16$$

where the quantizer output is taken to range over the $2N$ possible states denoted

$$\pm 1, \ \pm 2, \ \ldots, \ \pm N.$$

The step size adaption algorithm works as follows. If the previous output sample value was in the inner half of the quantizer range ($\pm 1, \ \pm 2, \ \ldots \pm N/2$), the step

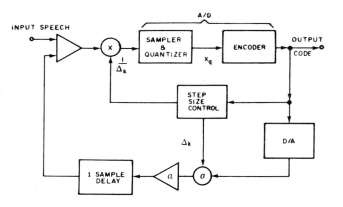

(a) Encoder

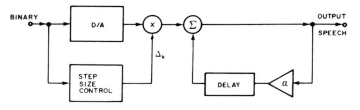

(b) Decoder

Figure 6.24   Typical ADPCM codec.

size decreases. This is intended to reduce the quantization noise. Otherwise, the step size increases. The step size Δ_{k+1} for sample $k+1$, is related to the previous step size Δ_k by

$$\Delta_{k+1} = M(|X_q(kT_s)|) \, \Delta_k^\beta \tag{6.54}$$

where

$|X_q(kT_s)|$ = the modulus of the quantizer output level for sample k
β = 0.98 is a ''leakage'' term that ensures that encoder and decoder step sizes converge after a transmission error, and

$$M = \begin{cases} 0.7718 & \text{for } |X_q(kT_s)| \le N/2 \\ 1.995 & \text{for } N/2 < |X_q(kT_s)| \le N. \end{cases}$$

The factor M is known as the step-size multiplier. For details on the performance of this ADPCM system with various levels of channel errors, see Goodman and Nash (8).

Adaptive predictive techniques attempt to minimize the rms levels of the difference signal (and the step size) relative to that of the original input. If the prediction is based on the past 10 samples or so of speech, it is generally referred to as short-time prediction. Long-time prediction using correlations from voiced period to voiced period can be utilized to further dynamically reduce this differential.

6.4.6 Adaptive Predictive Coding (APC)

The technique of Adaptive Predictive Coding (APC) is illustrated in Figure 6.25. This diagram shows the use of short-time and long-time predictors and a step-size adaption unit. In addition, the diagram shows a third feedback network around the quantizer. This provides quantization noise feedback to provide a noise shaping capability that takes into account the auditory masking properties of speech perception. This scheme represents an algorithm of relatively high complexity, but with the capability of providing good speech quality at bit rates down to approximately 12–16 kbit/s. Further details are given in Crochiere (4), Flanagan (6), and Jayant (12).

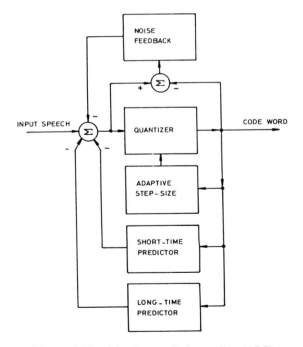

Figure 6.25 Adaptive predictive coding (APC).

6.4.7   Other Waveform Coding Techniques

Considerable research effort is currently being directed towards the study of waveform coding techniques. The aim is to obtain maximum performance quality and robustness against channel errors. This must be balanced against the desire to minimize complexity and coding delays because of the number of computations involved.

Research into methods of encoding telephone speech signals into bit rates at or below 16 kbit/s have focused on methods of efficiently encoding the prediction difference signal. They include methods such as tree encoding, vector quantizing, and other related techniques.

Another class of waveform coding techniques under study is known as *adaptive transform coding*. This involves segmenting the input speech waveform into short time blocks and then transforming each into the frequency domain. One of several versions of the Fast Fourier Transform (FFT) can be used for this purpose. The output frequency domain quantities from the discrete Fourier transform are then quantized and encoded. For details, see Seidl (9), Crochiere (4), Buzo (7) and *IEEE Trans. on Comms.* (8). For a tutorial review of discrete transform techniques and the FFT, see for example, (10).

6.5   PROBLEMS

6.1   A signal is given by

$$x(t) = \frac{\sin^2(\pi Bt)}{(\pi Bt)^2}$$

(1) Sketch $x(t)$ and its spectrum $X(f)$ for the case where $B = 1$ kHz.
(2) Assume that $x(t)$ is ideally sampled by being multiplied by an impulse train with sampling frequency $f_s = 4$ kHz. Sketch the spectrum of the sampled signal $x_s(t)$.
(3) Repeat (2) for the case where $f_s = 1.5$ kHz.

6.2   Consider the sampled signal $x_s(t)$ in Problem 6.1. If this is passed through a reconstruction filter, which is ideal low-pass with cut off frequency of 2 kHz, what will be the output waveform?

6.3   A signal

$$x(t) = 3 \cos 2\pi f_1 t + 2 \cos 3\pi f_1 t$$

is ideally sampled at $f_s = 8$ kHz. The sampled signal is passed through a reconstruction filter, which is ideal low-pass with cut off frequency of 4 kHz.

(1) Find the reconstruction filter output if $f_1 = 2$ kHz.

(2) Repeat for $f_1 = 3$ kHz.

6.4 Repeat Problem 6.3 for the case where the sampled signal consists of a sequence of flat-top pulses of width $\tau = 50$ μs and is of the form given by Equations (6.12) and (6.13).

6.5 Consider the low-pass filter in the figure below. Draw the circuit of an equivalent switched-capacitor filter.

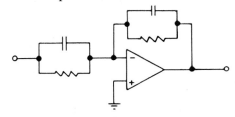

6.6 Using your switched capacitor filter circuit obtained in Problem 6.5, design a low-pass filter with a 3 dB bandwidth of 3.3 kHz and a low frequency gain of 40 dB. How would your filter perform as a PCM reconstruction filter?

6.7 The sinusoidal signal

$$x(t) = 2 \cos 2000\pi t$$

is to be transmitted in a PCM system in which the sample rate is $f_s = 8$ kHz, one sampling point occurring at time $t = 0$. Linear quantization is used with 16 levels over the input voltage range $(-2,2)$ volts.

Prepare a list showing code words representing the first eight samples of $x(t)$. State the assumptions made concerning the code words assigned to each quantization level.

6.8 An analogue signal is sampled, quantized, and transmitted using a PCM system. If each signal sample must be known at the receiving end to within 1 percent of the peak–peak full scale value, how many binary symbols must each transmitted digital word contain?

6.9 A color television signal is sampled at $f_s = 13.3$ MHz and PCM encoded with uniform quantization.

(1) If the signal to quantization noise ratio is to be greater than 30 dB, what bit rate is required?

(2) What is the appropriate allowable bandwidth of the television signal if the sampling rate to Nyquist rate ratio is comparable to that used for PCM voice encoding?

6.10 What is the dynamic range of a uniform PCM encoder with eight bits per sample and a minimum signal to quantization noise ratio of 34 dB?

6.11 Consider a PCM encoder using the A-law companding characteristic as illustrated in Figure 6.18(a). If the encoder is capable of encoding a maximum peak value of 500 mV, determine the codeword generated for a sample value of 100 mV.

6.12 Consider the μ-law companding characteristic given by Equation (6.33).

(1) Sketch the compression characteristic for $\mu = 0, 5, 10, 255$.

(2) Sketch the corresponding expandor characteristics.

6.13 The mean-squared quantization noise for the μ-law compressor characteristic with M levels is given approximately by

$$N = [\ln(1+\mu)]^2 \, (1 + \mu^2\sigma_x^2 + 2\mu E[|x|])/3M^2 \text{ (volts}^2)$$

where the input signal x has probability density function $f(x)$ concentrated over the range $(-1,1)$. Also,

$$\sigma_x^2 = \int_{-1}^{1} x^2 f(x) \, dx$$

and

$$E[|x|] = 2\int_{0}^{1} x \, f(x) \, dx.$$

(1) Show that $2E[|x|]/\sigma_x^2 = 1.414$ for the case where x is Gaussian (See Equation (4.30)).

(2) Show that $2E[|x|]/\sigma_x^2 = 1.6$ for the case where x is Laplacian. (Refer to Equation (4.31)).

(3) For each type of signal statistics compute the signal to quantization noise ratio

$$\text{SNR}_Q = 10 \log_{10}[\sigma_x^2/N]$$

for a 7-bit PCM μ-law coder with $\mu = 255$ for input signal power σ_x^2 equal to the maximum input signal.

(4) Repeat (3) for the case where the input signal power σ_x^2 is 40 dB below the maximum input signal.

6.14 A speech signal with maximum frequency component of 3 kHz is to be transmitted using delta modulation with sampling rate $f_s = 32$ kHz.

(1) Discuss the choice of appropriate step size Δ.

(2) Discuss the benefits of changing to a 64 kbit/s transmission rate.

6.6 REFERENCES

1. M. D. Paez, and T. H. Glisson, "Minimum mean-square-error quantization in speech PCM and DPCM systems" *IEEE Trans. Com.*, Vol. COM-20, pp. 255–230, April 1972.

2. L. Schweizer, "Planning aspects of quantizing distortion in telephone networks," *Telecommunication Journal*, Vol. 48, No. 1, 1981.

3. J. L. Flanagan, *Speech Analysis, Synthesis, and Perception*, Second Edition, New York, Springer-Verlag, 1972.

4. R. E. Crochiere, and J. L. Flanagan, "Current Perspectives in Digital Speech," *IEEE Communications Magazine*, Vol. 21, No. 1, pp. 32–40, Jan. 1983.

5. N. S. Jayant, "Digital coding of speech waveforms: PCM, DPCM, and DM quantizers," *Proc. IEEE*, Vol. 62, pp. 611–632, May 1974.

6. J. L. Flanagan, M. R. Schroeder, B. S. Atal, R. E. Crochiere, N. S. Jayant, and J. M. Tribolet, "Speech Coding," *IEEE Trans. Comm.*, Vol. COM-27, No. 4, pp. 710–737, April 1979.

7. A. Buzo, A. H. Gray, R. M. Gray, and J. D. Martiel, "Speech Coding Based on Vector Quantization," *IEEE Trans. ASSP*, Vol. 28, No. 5, pp. 562–574, October 1980.

8. *IEEE Transactions on Communications, Special issue on Bit Rate Reduction and Speech Interpolation*, Vol. COM-30, No. 4, April 1982.

9. R. A. Seidl, "A Tutorial Paper on Medium Bit Rate Speech Coding Techniques," *Australian Telecom. Research*, Vol. 17, No. 1, pp. 61–72, 1983.

10. M. J. Miller, "Discrete Signals and Frequency Spectra," *Handbook of Measurement Science*, Chapter 5, P. H. Sydenham (Ed.), John Wiley, 1982.

11. P. Bylanski and D. G. W. Ingram, *Digital Transmission Systems*, Peter Peregrinus Ltd., 1979.

12. N. S. Jayant, "Coding speech at low bit rates," *IEEE Spectrum*, Vol. 23, No. 8, pp. 58–63, August 1986.

INDEX

263